ABC OF
CLINICAL GENETICS,
THIRD EDITION

ry

.edical Association   BMA

ook constitute   rc tance f he conditions set out b

ABC OF
CLINICAL GENETICS

Third edition

Helen M Kingston

Consultant Clinical Geneticist, Regional Genetic Service,
St Mary's Hospital, Manchester, UK

First published 1989
Second impression (revised) 1990
Second edition 1994
Second impression (revised) 1997
Third impression 1999
Third edition 2002
by BMJ Books, BMA House, Tavistock Square,
London WC1H 9JR

www.bmjbooks.com

Cover image depicts a computer representation of
the beta DNA molecule. Produced with permission
from Prof K Seddon and Dr T Evans,
Queen's University, Belfast/Science Photo Library.

British Library Cataloguing in Publication Data
A catalogue record for this book is available from the British Library

ISBN 0-7279-1627-0

Typeset by Newgen Imaging Systems (P) Ltd., Chennai, India

Printed in Malaysia by Times Offset

Contents

Contributors

David Gokhale
Scientist, Molecular Genetic Laboratory, Regional Genetic Service, St Mary's Hospital, Manchester

Lauren Kerzin-Sturrar
Principal Genetic Associate, Regional Genetic Service, St Mary's Hospital, Manchester

Tara Clancy
Senior Genetic Associate, Regional Genetic Service, St Mary's Hospital, Manchester

Bronwyn Kerr
Consultant Clinical Geneticist, Regional Genetic Service, St Mary's Hospital, Manchester

Preface

Since the first edition of this book in 1989 there have been enormous changes in clinical genetics, reflecting the knowledge generated from the tremendous advances in molecular biology, culminating in the publication of the first draft of the human genome sequence in 2001, and the dissemination of information via the internet. The principles of genetic assessment and the aims of genetic counselling have not changed, but the classification of genetic disease and the practice of clinical genetics has been significantly altered by this new knowledge. To interpret all the information now available it is necessary to understand the basic principles of inheritance and its chromosomal and molecular basis. Recent advances in medical genetics have had a considerable impact on other specialties, providing a new range of molecular diagnostic tests applicable to many branches of medicine, and more patients are presenting to their general practitioners with concerns about a family history of disorders such as cancer. Increasingly, other specialties are involved in the genetic aspects of the conditions they treat and need to provide information about genetic risk, undertake genetic testing and provide appropriate counselling. All medical students, irrespective of their eventual career choice therefore need to be familiar with genetic principles, both scientific and clinical, and to be aware of the ethical implications of genetic technologies that enable manipulation of the human genome that may have future application in areas such as gene therapy of human cloning. The aim of this third edition of the ABC is therefore to provide an introduction to the various aspects of medical genetics for medical students, clinicians, nurses and allied professionals who are not working within the field of genetics, to generate an interest in the subject and to guide readers in the direction of further, more detailed information.

In producing this edition, the chapters on molecular genetics and its application to clinical practice have been completely re-written, bringing the reader up to date with current molecular genetic techniques and tests as they are applied to inherited disorders. An introduction to the internet in human genetics has also been included. There are new chapters on genetic services, genetic assessment and genetic counselling together with a new chapter highlighting the clinical and genetic aspects of some of the more common single gene disorders. Substantial alterations have been made to most other chapters so that they reflect current practice and knowledge, although some sections of the previous edition remain. A glossary of terms is included for readers who are not familiar with genetic terminology, a further reading list is incorporated and a list of websites included to enable access to data that is changing on a daily basis. As in previous editions, illustrations are a crucial component of the book, helping to present complex genetic mechanisms in an easily understood manner, providing photographs of clinical disorders, tabulating genetic diseases too numerous to be discussed individually in the text and showing the actual results of cytogenetic and molecular tests.

I am grateful to many colleagues who have helped me in producing this edition of the ABC. In particular, I am indebted to Dr David Gokhale who has re-written chapters 17, 18 and 20, and has provided the majority of the illustrations for chapters 16, 17 and 18. I am also grateful to Lauren Kerzin-Storrar and Tara Clancy for writing chapter 3 and to Dr Bronwyn Kerr for contributing to chapter 11. Numerous colleagues have provided illustrations and are acknowledged throughout the book. In particular, I would like to thank Professor Dian Donnai, Dr Lorraine Gaunt and Dr Sylvia Rimmer who have provided many illustrations for this as well as previous editions, and to Helena Elliott who has prepared most of the cytogenetic pictures incorporated into this new edition. I am also very grateful to the families who allowed me to publish the clinical photographs that are included in this book to aid syndrome recognition.

Helen M Kingston

1 Clinical genetic services

Development of medical genetics

The speciality of medical genetics is concerned with the study of human biological variation and its relationship to health and disease. It encompasses mechanisms of inheritance, cytogenetics, molecular genetics and biochemical genetics as well as formal, statistical and population genetics. Clinical genetics is the branch of the specialty involved with the diagnosis and management of genetic disorders affecting individuals and their families.

Genetic counselling clinics were first established in the USA in 1941 and in the UK in 1946. Some of the disorders dealt with in these early clinics were ones that are seldom referred today, such as skin colour, eye colour, twinning and rhesus haemolytic disease. Other referrals were very similar to those being seen today – namely, mental retardation, neural tube defects and Huntington disease. Prior to the inception of these clinics, the patterns of dominant and recessive inheritance, described by Mendel in 1865, were recognised in human disorders. Autosomal recessive inheritance of alkaptonuria had been recognised in 1902 by Archibald Garrod, who also introduced the term "inborn errors of metabolism". In 1908, the Hardy–Weinberg principle of population genetics was delineated and remains the basis of calculating carrier frequencies for autosomal recessive disorders. The term, "genetic counselling" was introduced by Sheldon Reed, whose definition of the process is given later in this chapter.

DNA, initially called "nuclein", had been discovered by Meischer in 1867 and the first illustration of human chromosomes was published by Walther Fleming in 1882 although the term "chromosome" was not coined until 1888 and the chromosomal basis of mendelism only proposed in 1903. The correct chromosome number in humans was not established until 1956 and the association between trisomy 21 and Down syndrome was reported in 1959. The structure of DNA was determined by Watson and Crick in 1953 and by 1966 the complete genetic code had been cracked. DNA analysis became possible during the 1970s with the discovery of restriction endonucleases and development of the Southern blotting technique. These advances have led to the mapping and isolation of many genes and subsequent mutation analysis. Enormous advances in molecular biology techniques have resulted in publication of the draft sequence of the human genome in 2000. As a result of these scientific discoveries and developments, clinical geneticists are able to use chromosomal analysis and molecular genetic tests to diagnose many genetic disorders.

Genetic disease

Genetic disorders place considerable health and economic burdens not only on affected individuals and their families but also on the community. As more environmental diseases are successfully controlled those that are wholly or partly genetically determined are becoming more important. Despite a general fall in the perinatal mortality rate, the incidence of lethal malformations in newborn infants remains constant. Between 2 and 5% of all liveborn infants have genetic disorders or congenital malformations. These disorders have been estimated to account for one third of admissions to paediatric wards, and they contribute appreciably to perinatal and childhood mortality. Many common diseases in adult life also

Figure 1.1 Gregor Mendel 1822–84

Figure 1.2 Archibald Garrod 1858–1936

Figure 1.3 The discoverers of the structure of DNA. James Watson (b. 1928) at left and Francis Crick (b. 1916), seen with their model of part of a DNA molecule in 1953 (with permission from A Barrington Brown/Science Photo Library)

Table 1.1 Prevalence of genetic disease

Type of genetic disease	Estimated prevalence per 1000 population
Single gene	
Autosomal dominant	2–10
Autosomal recessive	2
X linked recessive	1–2
Chromosomal abnormalities	6–7
Common disorders with appreciable genetic component	7–10
Congenital malformations	20
Total	38–51

have a considerable genetic predisposition, including coronary heart disease, diabetes and cancer.

Though diseases of wholly genetic origin are individually rare, they are numerous (several thousand single gene disorders are described) and are therefore important. Genetic disorders are incurable and often severe. Some are amenable to treatment but many are not, so that the emphasis is often placed on prevention, either of recurrence within an affected family, or of complications in a person who is already affected.

Increasing awareness, both within the medical profession and in the general population, of the genetic contribution to disease and the potential implications of a positive family history, has led to an increasing demand for specialist clinical genetic services. Some aspects of genetics are well established and do not require referral to a specialist genetics clinic – for example, the provision of amniocentesis to exclude Down syndrome in pregnancies at increased risk. Other aspects are less well understood – for example the application of molecular genetic analysis in clinical practice, which is an area of rapidly advancing technology requiring the facilities of a specialised genetics centre.

Organisation of genetic services

In the UK, NHS genetic services are provided in integrated regional centres based in teaching hospitals, incorporating clinical and laboratory departments usually in close liaison with academic departments of human genetics.

Clinical genetics

Clinical services are provided by consultant clinical geneticists, specialist registrars and genetic associates (nurses or graduates with specialist training in genetics and counselling). Most clinical genetic departments provide a "hub and spoke" service, undertaking clinics in district hospitals as well as at the regional centre. Patients referred to the genetic clinic are contacted initially by the genetic associate and many are visited at home before attending the clinic. The purpose of the home visit is to explain the nature of the genetic clinic appointment, determine the issues of importance to the family and obtain relevant family history information. The genetic associate is usually present at the clinic appointment and participates in the counselling process with the clinical geneticist. At the clinic appointment genetic investigations may be instituted to establish or confirm a diagnosis and information is given to the individual or family about the condition regarding diagnosis, prognosis, investigation, management and genetic consequences. Written information is usually provided after the clinic appointment so that the family have a record of the various aspects discussed. After the appointment, follow-up visits at home or in the clinic are arranged as necessary. The genetic associate plays an important role in liaising with primary care and other agencies involved with the family.

There are a wide variety of reasons leading to referral to the genetic clinic. The referral may be for diagnosis in cases where a genetic disorder is suspected; for counselling when a genetic condition has been identified; for genetic investigation of family members when there is a family history of an inherited disorder; or for information regarding prenatal diagnosis. The disorders seen include sporadic birth defects and chromosomal syndromes as well as mendelian, mitochondrial and multifactorial conditions. Specialist or multidisciplinary clinics are provided by some genetic centres, such as for dysmorphology, inherited cancers, neuromuscular disorders, Huntington disease, Marfan syndrome, ophthalmic disorders or hereditary deafness.

Box 1.1 Type of genetic disease

Single gene (mendelian)
- Numerous though individually rare
- Clear pattern of inheritance
- High risk to relatives

Multifactorial
- Common disorders
- No clear pattern of inheritance
- Low or moderate risk to relatives

Chromosomal
- Mostly rare
- No clear pattern of inheritance
- Usually low risk to relatives

Somatic mutation
- Accounts for mosaicism
- Cause of neoplasia

Figure 1.4 Explaining genetic mechanisms and risks during genetic counselling

Box 1.2 Common reasons for referral to a genetic clinic
- Children with congenital abnormalities (birth defects), learning disability, dysmorphic features
- Children with chromosomal disorders or inherited conditions
- Adults affected by congenital abnormality or an inherited condition
- Adults known to carry or at risk of carrying, a balanced chromosomal rearrangement
- Couples who have lost a child or stillborn baby with a congenital abnormality or inherited condition
- Couples who have suffered reproductive loss (termination of pregnancy for fetal abnormality or recurrent miscarriage)
- Pregnant women and their partners, when fetal abnormality is detected
- Children and adults with a family history of a known genetic disorder
- Adults at risk of developing an inherited condition who may request predictive testing
- Couples who may transmit a genetic condition to their children
- Individuals with a family history of a common condition with a strong genetic component, including familial cancers

Information about clinical genetic centres in the UK can be obtained from the British Society for Human Genetics (incorporating the Clinical Genetics Society and the Association of Genetic Nurses) website at www.bshg.org.uk

Cytogenetics

Cytogenetic laboratories undertake chromosomal analysis on a variety of samples including whole blood (collected into lithium heparin), amniotic fluid, chorion villus or placental samples, cultures of solid tissues and bone marrow aspirates. Analysis is undertaken to diagnose chromosomal disorders when a diagnosis is suspected clinically, to identify carriers of familial chromosomal rearrangements when there is a family history and to provide information related to therapy and prognosis in certain neoplastic conditions. Some of the main indications for performing chromosomal analysis are listed in the box.

Routine chromosomal analysis requires the study of metaphase chromosomes in cultured cells. Results are usually available in 1–3 weeks. Molecular genetic analysis by fluorescence in situ hydridisation (FISH) studies is possible for certain conditions. These studies are usually performed on cultured cells, but in some cases (such as urgent prenatal confirmation of trisomy 21) rapid results may be obtained by analysis of interphase nuclei in uncultured cells.

Information about cytogenetic centres in the UK can be obtained from the Association of Clinical Cytogeneticists (ACC) website at www.acc.org.uk

Molecular genetics

Molecular genetic laboratories offer a range of DNA tests. Direct mutation analysis is available for certain conditions and provides confirmation of clinical diagnosis in affected individuals, presymptomatic diagnosis for individuals at risk of specific conditions, carrier detection and prenatal diagnosis. Mutation analysis for rare disorders is usually undertaken on a supra-regional or national basis in designated laboratories. For mendelian disorders in which mutation analysis is not possible, gene tracking using linked DNA markers may be used to predict inheritance of certain conditions (for example Marfan syndrome and Neurofibromatosis type 1) if the family structure is suitable.

DNA can be extracted from any tissue containing nucleated cells, including stored tissue blocks. Tests are usually performed on whole blood collected into EDTA anticoagulant, or buccal samples obtained by scraping the inside of the cheek or by mouth wash. Once extracted, frozen DNA samples can be stored indefinitely. Samples can therefore be collected from family members and stored for future analysis of disorders that are currently not amenable to molecular analysis.

In the UK, the Clinical Molecular Genetics Society (CMGS) provides data on molecular services offered by individual laboratories through their website at www.cmgs.org.uk

Biochemical genetics

Specialised biochemical genetic departments offer clinical and laboratory services for a range of inherited metabolic disorders. Routine neonatal screening for conditions such as phenylketonuria (PKU) and congenital hypothyroidism are undertaken on neonatal blood samples taken from all newborn babies. Investigations performed on children presenting with other metabolic disorders are carried out on a range of samples including urine, blood, CSF, cultured fibroblasts and muscle biopsies. Tests are undertaken to identify conditions such as disorders of amino acids, organic acids and mucopolysaccharides, lysosomal and lipid storage diseases, and

---

Box 1.3 Common reasons for cytogenetic analysis

Postnatal
- Newborn infants with birth defect
- Children with learning disability
- Children with dysmorphic features
- Familial chromosomal rearrangement in relative
- Infertility
- Recurrent miscarriages

Prenatal
- Abnormalities on ultrasound scan
- Increased risk of Down syndrome (maternal age or biochemical screening)
- Previous child with a chromosomal abnormality
- One parent carries a structural chromosomal abnormality

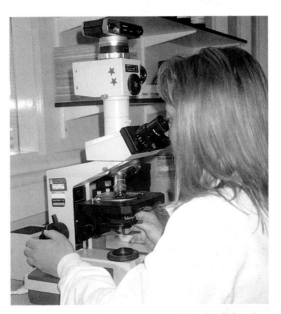

Figure 1.5 Conventional cytogenic analysis using light microscopy

Box 1.4 Some common reasons for molecular genetic analysis

- Cystic fibrosis
- Haemoglobinopathies
- Duchenne and Becker muscular dystrophy
- Myotonic dystrophy
- Huntington disease
- Fragile X syndrome
- Spinal muscular atrophy
- Spinocerebellar ataxia
- Hereditary neuropathy (Charcot-Marie-Tooth)
- Familial breast cancer (*BRCA 1* and *2*)
- Familial adenomatous polyposis

Figure 1.6 Typical molecular genetics laboratory

peroxisomal and mitochondrial disorders. Tests for other metabolites or enzymes are performed when a diagnosis of a specific disorder is being considered. In the UK, the Society for the Study of Inborn Errors of Metabolism publishes information on centres providing biochemical genetic tests. Their website address is www.ssiem.org.uk

Genetic registers

Genetic registers have been in use in the UK for about 30 years and most genetic centres operate some type of register for specified disorders. In some cases the register functions as a reference list of cases for diagnostic information, but generally the system is used to facilitate patient management. Ascertainment of cases is usually through referrals made to the genetic centre. Less often there is an attempt to actively ascertain all affected cases within a given population. To function effectively most registers contain information about relatives at risk as well as affected individuals and may contain information from genetic test results. Establishment of a register enables long-term follow up of family members. This is important for children at risk who may not need counselling and investigation for many years. A unique aspect of a family based genetic register is that it includes clinically unaffected individuals who may require continued surveillance and enables continued contact with couples at risk of transmitting disorders to their children.

Registers are particularly useful for disorders amenable to DNA analysis in which advances of clinical importance are likely to improve future genetic testing and where families will need to be updated with new information. Disorders suited to a register approach include dominant disorders with late onset such as Huntington disease and myotonic dystrophy where pre-symptomatic diagnosis may be requested by some family members or health surveillance is needed by affected individuals; X linked disorders such as Duchenne and Becker muscular dystrophy where carrier testing is offered to female relatives, and chromosomal translocations where relatives benefit from carrier testing. Registers can also provide data on the incidence and natural course of disease as well as being used to monitor the provision and effectiveness of services. Genetic register information is held on computer and is subject to the Data Protection Act. No one has his/her details included on a register without giving informed consent.

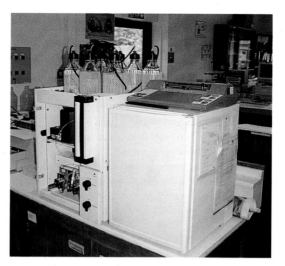

Figure 1.7 Amino acid analyser in biochemical genetics laboratory

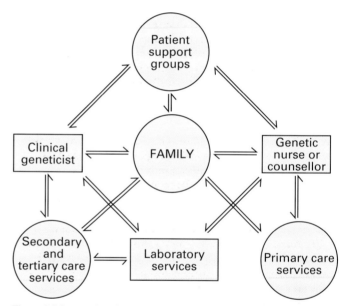

Figure 1.8 Interactions between families with genetic disorders and various medical and support services

2 Genetic assessment

Genetic diagnosis

The role of clinical geneticists is to establish an accurate diagnosis on which to base counselling and then to provide information about prognosis and follow up, the risk of developing or transmitting the disorder, and the ways in which this may be prevented or ameliorated. Throughout, the family requires support in adjusting to the implications of genetic disease and the consequent decisions that may have to be made.

History taking

Diagnosis of genetic disorders is based on taking an accurate history and performing clinical examination, as in any other branch of medicine. The history and examination will focus on aspects relevant to the presenting complaint. When a child presents with birth defects, for example, information needs to be gathered concerning parental age, maternal health, pregnancy complications, exposure to potential teratogens, fetal growth and movement, prenatal ultrasound scan findings, mode of delivery and previous pregnancy outcomes. Information regarding similar or associated abnormalities present in other family members should also be sought. In conditions with onset in adult life, the age at onset, mode of presentation and course of the disease in affected relatives should be documented, together with the ages reached by unaffected relatives.

Examination

Thorough physical examination is required, but emphasis will be focused on relevant anatomical regions or body systems. Detailed examination of children with birth defects or dysmorphic syndromes is crucial in attempting to reach a diagnosis. A careful search should be made for both minor and major congenital abnormalities. Measurements of height, weight and head circumference are important and standard growth charts and tables are available for a number of specific conditions, such as Down syndrome, Marfan syndrome and achondroplasia. Other measurements, including those of body proportion and facial parameters may be appropriate and examination findings are often best documented by clinical photography. In some cases, clinical geneticists will need to rely on the clinical findings of other specialists such as ophthalmologists, neurologists and cardiologists to complete the clinical evaluation of the patient.

The person attending the clinic may not be affected, but may be concerned to know whether he or she might develop a particular disorder or transmit it to any future children. In such cases, the diagnosis in the affected relative needs to be clarified, either by examination or by review of relevant hospital records (with appropriate consent). Apparently unaffected relatives should be examined carefully for minor or early manifestations of a condition to avoid inappropriate reassurance. In myotonic dystrophy, for example, myotonia of grip and mild weakness of facial muscles, sterno-mastoids and distal muscles may be demonstrated in asymptomatic young adults and indicate that they are affected. Subjects who may show signs of a late onset disorder should be examined before any predictive genetic tests are done, so that the expectation of the likely result is realistic. Some young adults who request predictive tests to reassure themselves that they are not affected may not wish to proceed with definitive tests if they are told that their clinical examination is not entirely normal.

Figure 2.1 Recording family history details by drawing a pedigree

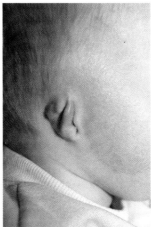 

Figure 2.2 The presence of one congenital anomaly should prompt a careful search for other anomalies

Figure 2.3 Physical measurements are an important part of clinical examination

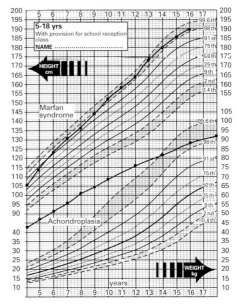

Figure 2.4 Growth chart showing typical heights in Marfan syndrome and Achondroplasia compared to normal centiles

Investigations

Investigation of affected individuals and family members may include conventional tests such as x-rays and biochemical analysis as well as cytogenetic and molecular genetic tests. A search for associated anomalies in children with chromosomal disorders often includes cranial, cardiac and renal imaging along with tests for other specific components of the particular syndrome, such as immune deficiency. In some genetic disorders affected individuals may require regular investigations to detect disease-associated complications, such as cardiac arrhythmias and reduced lung function in myotonic dystrophy. Screening for disease complications in asymptomatic relatives at risk of a genetic disorder may also be appropriate, for example, 24-hour urine catecholamine estimation and abdominal scans for individuals at risk of von Hippel–Lindau disease.

Drawing a pedigree

Accurate documentation of the family history is an essential part of genetic assessment. Family pedigrees are drawn up and relevant medical information on relatives sought. There is some variation in the symbols used for drawing pedigrees. Some suggested symbols are shown in the figure. It is important to record full names and dates of birth of relatives on the pedigree, so that appropriate hospital records can be obtained if necessary. Age at onset and symptoms in affected relatives should be documented. Specific questions should be asked about abortions, stillbirth, infant death, multiple marriages and consanguinity as this information may not always be volunteered.

When a pedigree is drawn, it is usually easiest to start with the person seeking advice (the consultand). Details of first degree relatives (parents, siblings and children) and then second degree relatives (grandparents, aunts, uncles, nieces and nephews) are added. If indicated, details of third degree relatives can be added. If the consultand has a partner, a similar pedigree is constructed for his or her side of the family. The affected person (proband) through whom the family has been ascertained is usually indicated by an arrow.

Confirmation of a clinical diagnosis may identify a defined mode of inheritance for some conditions. In others, similar phenotypes may be due to different underlying mechanisms, for example, limb girdle muscular dystrophy may follow dominant or recessive inheritance and the pedigree may give clues as to which mechanism is more likely. In cases where no clinical diagnosis can be reached, information on genetic risk can be given if the pedigree clearly indicates a particular mode of inheritance. However, when there is only a single affected individual in the family, recurrence risk is difficult to quantify if a clinical diagnosis cannot be reached.

Estimation of risk

For single gene disorders amenable to mutation analysis, risks to individuals of developing or transmitting particular conditions can often be identified in absolute terms. In many conditions, however, risks are expressed in terms of probabilities calculated from pedigree data or based on empirical risk figures. An important component of genetic counselling is explaining these risks to families in a manner that they can understand and use in decision making.

Mendelian disorders due to mutant genes generally carry high risks of recurrence whereas chromosomal disorders generally have a low recurrence risk. For many common conditions there is no clearly defined pattern of inheritance

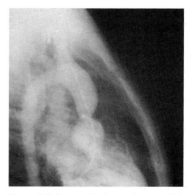

Figure 2.5 Supravalvular aortic stenosis in a child with William syndrome

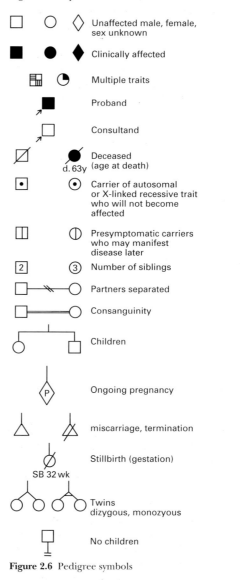

Figure 2.6 Pedigree symbols

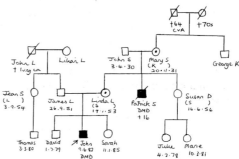

Figure 2.7 Hand drawn pedigree of a family with Duchenne muscular dystrophy identifying obligate carriers and other female relatives at risk

and the empirical figures for risk of recurrence are based on information derived from family studies. There is considerable heterogeneity observed in many genetic disorders. Similar phenotypes may be due to mutations at different loci (locus heterogeneity) or to different modes of inheritance. In autosomal recessive deafness there is considerable locus heterogeneity with over 30 different loci known to cause non-syndromic severe congenital deafness. The risk to offspring of two affected parents will be 100% if their deafness is due to gene mutations at the same locus, but negligible if due to gene mutations at different loci. In some disorders, for example hereditary spastic paraplegia and retinitis pigmentosa, autosomal dominant, autosomal recessive and X linked recessive inheritance have been documented. Definite recurrence risks cannot be given if there is only one affected person in the family, since dominant and recessive forms cannot be distinguished clinically.

Perception of risk is affected by the severity of the disorder, its prognosis and the availability of treatment or palliation. All these aspects need to be considered when information is given to individuals and families. The decisions that couples make about pregnancy are influenced partly by the risk of transmitting the disorder, and partly by its severity and the availability of prenatal diagnosis. A high risk of a mild or treatable disorder may be accepted, whereas a low risk of a severe disorder can have a greater impact on reproductive decisions. Conversely, where no prenatal diagnosis is possible, a high risk may be more acceptable for a lethal disorder than for one where prolonged survival with severe handicap is expected. Moral and religious convictions play an important role in an individual's decision making regarding reproductive options and these beliefs must be respected.

Consanguinity

Consanguinity is an important issue to identify in genetic assessment because of the increased risk of autosomal recessive disorders occurring in the offspring of consanguineous couples. Everyone probably carries at least one harmful autosomal recessive gene. In marriages between first cousins the chance of a child inheriting the same recessive gene from both parents that originated from one of the common grandparents is 1 in 64. A different recessive gene may similarly be transmitted from the other common grandparent, so that the risk of homozygosity for a recessive disorder in the child is 1 in 32. If everyone carries two recessive genes, the risk would be 1 in 16.

Marriage between first cousins generally increases the risk of severe abnormality and mortality in offspring by 3–5% compared with that in the general population. The increased risk associated with marriage between second cousins is around 1%. Marriage between first and second degree relatives is almost universally illegal, although marriages between uncles and nieces occur in some Asian countries. Marriage between third degree relatives (between cousins or half uncles and nieces) is more common and permitted by law in many countries.

The offspring of incestuous relationships are at high risk of severe abnormality, mental retardation and childhood death. Only about half of the children born to couples who are first degree relatives are normal and this has important implications for decisions about termination of pregnancy or subsequent adoption.

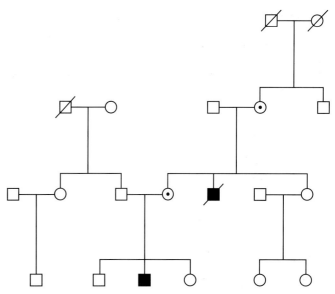

Figure 2.8 The same pedigree as Figure 2.7 drawn using Cyrillic computer software (Cherwell Scientific Publishing)

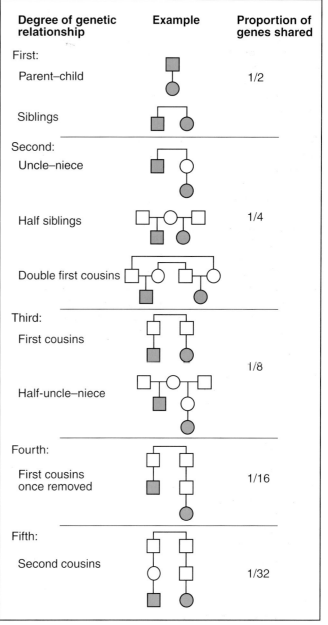

Figure 2.9 Proportion of genes shared by different relatives

3   Genetic counselling

Genetic counselling has been defined as a communication process with both educative and psychotherapeutic aims. While genetic counselling must be based on accurate diagnosis and risk assessment, its use by patients and families will depend upon the way in which the information is given and its psychosocial impact addressed. The ultimate aim of genetic counselling is to help families at increased genetic risk to live and reproduce as normally as possible.

While genetic counselling is a comprehensive activity, the particular focus will depend upon the family situation. A pregnant couple at high genetic risk may need to make urgent decisions concerning prenatal diagnosis; parents of a newly diagnosed child with a rare genetic disorder may be desperate for further prognostic information, while still coming to terms with the diagnosis; a young adult at risk of a late onset degenerative disorder may be well informed about the condition, but require ongoing discussions about whether to go ahead with a presymptomatic test; and a teenage girl, whose brother has been affected with an X linked disorder, may be apprehensive to learn about the implication for her future children, and unsure how to discuss this with her boyfriend. Being able to establish the individual's and the family's particular agenda, to present information in a clear manner, and to address psychosocial issues are all crucial skills required in genetic counselling.

Psychosocial issues

The psychosocial impact of a genetic diagnosis for affected individuals and their families cannot be over emphasised. The diagnosis of any significant medical condition in a child or adult may have psychological, financial and social implications, but if the condition has a genetic basis a number of additional issues arise. These include guilt and blame, the impact on future reproductive decisions and the genetic implications to the extended family.

Guilt and blame

Feelings of guilt arise in relation to a genetic diagnosis in the family in many different situations. Parents very often express guilt at having transmitted a genetic disorder to their children, even when they had no previous knowledge of the risk. On the other hand, parents may also feel guilty for having taken the decision to terminate an affected pregnancy. Healthy members of a family may feel guilty that they have been more fortunate than their affected relatives and at-risk individuals may feel guilty about imposing a burden onto their partner and partner's family. Although in most situations the person expressing guilt will have played no objective causal role, it is important to allow him or her to express these concerns and for the counsellor to reinforce that this is a normal human reaction to the predicament.

Blame occurs perhaps less often than anticipated by families. Although parents often fear that their children will blame them for their adverse genetic inheritance, in practice this happens infrequently and usually only when the parents have knowingly withheld information about the genetic risk. Blame can sometimes occur in families where only one member of a couple carries the genetic risk ("It wasn't *our* side"), but

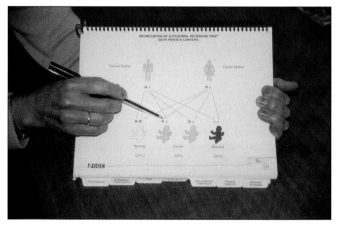

Figure 3.1 Explanation of genetic mechanisms is an important component of genetic counselling

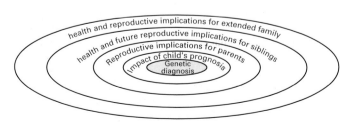

Figure 3.2 Impact of genetic diagnosis

again this is less likely to occur when the genetic situation has been explained and is understood.

Reproductive decision making

Couples aware of an increased genetic risk to their offspring must decide whether this knowledge will affect their plans for a family. Some couples may be faced with a perplexing range of options including different methods of prenatal diagnosis and the use of assisted reproductive technologies. For others the only available option will be to choose between taking the risk of having an affected child and remaining childless. Couples may need to reconsider these choices on repeated occasions during their reproductive years.

Most couples are able to make reproductive choices and this is facilitated through access to full information and counselling. Decision making may be more difficult in particular circumstances, including marital disagreement, religious or cultural conflict, and situations where the prognosis for an affected child is uncertain. For many genetic disorders with variable severity, although prenatal diagnosis can be offered, the clinical prognosis for the fetus cannot be predicted. When considering reproductive decisions, it can also be difficult for a couple to reconcile their love for an affected child or family member, with a desire to prevent the birth of a further affected child.

Impact on the extended family

The implications of a genetic diagnosis usually reverberate well beyond the affected individual and his or her nuclear family. For example, the parents of a boy just diagnosed with Duchenne muscular dystrophy will not only be coming to terms with his anticipated physical deterioration, but may have concerns that a younger son could be affected and that daughters could be carriers. They also face the need to discuss the possible family implications with the mother's sisters and female cousins who may already be having their own children. This is likely to be distressing even when family relationships are intact, but will be further complicated in families where relationships are less good.

Family support can be very important for people coping with the impact of a genetic disorder. When there are already several affected and carrier individuals in a family, the source of support from other family members can be compromised. For some families affected by disorders such as Huntington disease and familial breast cancer (*BRCA 1 and 2*), a family member in need of support may be reluctant to burden relatives who themselves are coping with the disease or fears about their own risks. They may also be hesitant to discuss decisions about predictive or prenatal testing with relatives who may have made different choices themselves. The need for an independent friend or counsellor in these situations is increased.

Bereavement

Bereavement issues arise frequently in genetic counselling sessions. These may pertain to losses that have occurred recently or in the past. A genetic disorder may lead to reproductive loss or death of a close family member. The grief experienced after termination of pregnancy following diagnosis of abnormality is like that of other bereavement reactions and may be made more intense by parents' feelings of guilt. After the birth of a baby with congenital malformations, parents mourn the loss of the imagined healthy child in addition to their sadness about their child's disabilities, and this chronic sorrow may be ongoing throughout the affected child's life.

Box 3.2 Possible reproductive options for those at increased genetic risk

Pregnancy without prenatal diagnosis
- Take the risk
- Limit family size

Pregnancy with prenatal diagnosis
- Chorion villus sampling
- Amniocentesis
- Ultrasound

Donor sperm
- Male partner has autosomal dominant disorder
- Male partner has chromosomal abnormality
- Both partners carriers for autosomal recessive disorder

Donor egg
- Female partner carrier for X linked disorder
- Female partner has autosomal dominant disorder
- Female partner has chromosomal abnormality
- Both partners carriers for autosomal recessive disorder

Preimplantation genetic diagnosis and IVF
- Available for a small number of disorders

Contraception
- Couples who chose to have no children
- Couples wanting to limit family size
- Couples waiting for new advances

Sterilisation
- Couples whose family is complete
- Couples who chose to have no children

Fostering and adoption
- Couples who want children, but find all the above options unacceptable

---

Box 3.3 Lay support groups

Contact a Family
Produces comprehensive directory of individual conditions and their support groups in the UK
http://www.cafamily.org.uk

Genetic Interest Group
Alliance of lay support groups in the UK which provides information and presents a unified voice for patients and families, in social and political forums
http://www.gig.org.uk

Antenatal Results and Choices (ARC)
Publishes and distributes an invaluable booklet for parents facing the decision whether or not to terminate a pregnancy after diagnosis of abnormality, and offers peer telephone support
http://www.cafamily.org.uk/Direct/f26html

Unique
Support group for individuals with rare chromosome disorders and their families
http://www.rarechromo.org

European Alliance of Genetic Support Groups
A federation of support groups in Europe helping families with genetic disorders
http://www.ghq-ch.com/eags

The Genetic Alliance
An umbrella organisation in the US representing individual support groups and aimed at helping all individuals and families with genetic disorders
http://www.geneticalliance.org

National Organisation for Rare Disorders (NORD)
A federation of voluntary groups in the US helping people with rare conditions
http://www.rarediseases.org

Long-term support

Many families will require ongoing information and support following the initial genetic counselling session, whether coping with an actual diagnosis or the continued risk of a genetic disorder. This is sometimes coordinated through regional family genetic register services, or may be requested by family members at important life events including pregnancy, onset of symptoms, or the death of an affected family member. Lay support groups are an important source of information and support. In addition to the value of contact with other families who have personal experience of the condition, several groups now offer the help of professional care advisors. In the UK there is an extensive network of support groups for a large number of individual inherited conditions and these are linked through two organisations: Contact a Family (www.cofamily.co.uk) and the Genetic Interest Group (GIG) (www.gig.org.uk).

Figure 3.3 Lay support group leaflets

Counselling around genetic testing

Genetic counselling is an integral part of the genetic testing process and is required because of the potential impact of a test result on an individual and family, as well as to ensure informed choice about undergoing genetic testing. The extent of the counselling and the issues to be addressed will depend upon the type of test being offered, which may be diagnostic, presymptomatic, carrier or prenatal testing.

Testing to confirm a clinical diagnosis

When a genetic test is requested to confirm a clinical diagnosis in a child or adult, specialist genetic counselling may not be requested until after the test result. It is therefore the responsibility of the clinician offering the test to inform the patient (or the parents, if a child is being tested) before the test is undertaken, that the results may have genetic as well as clinical implications. Confirming the diagnosis of a genetic disorder in a child, for example, may indicate that younger siblings are also at risk of developing the disorder. For late onset conditions such as Huntington disease, it is crucial that samples sent for diagnostic testing are from patients already symptomatic, as there are stringent counselling protocols for presymptomatic testing (see below).

Presymptomatic testing

Genetic testing in some late onset autosomal dominant disorders can be used to predict the future health of a well individual, sometimes many decades in advance of onset of symptoms. For some conditions, such as Huntington disease, having this knowledge does not currently alter medical management or prognosis, whereas for others, such as familial breast cancer, there are preventative options available. For adult onset disorders, testing is usually offered to individuals above the age of 18. For conditions where symptoms or preventative options occur in late childhood, such as familial adenomatous polyposis, children are involved in the testing decision. Presymptomatic testing is most commonly done for individuals at 50% risk of an autosomal dominant condition. Testing someone at 25% is avoided wherever possible, as this could disclose the status of the parent at 50% risk who may not want to have this cinformation. There are clear guidelines for provision of genetic counselling for presymptomatic testing, which include full discussion of the potential drawbacks of testing (psychological, impact on the family and financial), with ample opportunity for an individual to withdraw from testing right up until disclosure of results, and a clear plan for follow up.

Box 3.4 Genetic testing defined

Diagnostic – confirms a clinical diagnosis in a symptomatic individual

Presymptomatic ("predictive") – confirms that an individual will develop the condition later in life

Susceptibility – identifies an individual at increased risk of developing the condition later in life

Carrier – identifies a healthy individual at risk of having children affected by the condition

Prenatal – diagnoses an affected fetus

Genetic Counselling
- One or more sessions
- Molecular confirmation of diagnosis in affected relative
- Discussion of clinical and genetic aspects of condition, and impact on family

Patient requests test (interval of several months suggested)

Pre-test Counselling
- At least one session
- Seen by 2 members of staff (usually clinical geneticist and genetic counsellor)
- Involvement of partner encouraged
- Full discussion about:
  - Motivation for requesting test
  - Alternatives to having a test
  - Potential impact of test result
    - Psychological
    - Financial
    - Social (relationships with partner/family
  - Strategies for coping with result

Figure 3.4 continued

Carrier testing

Testing an individual to establish his or her carrier state for an autosomal or X linked recessive condition or chromosomal rearrangement, will usually be for future reproductive, rather than health, implications. Confirmation of carrier state may indicate a substantial risk of reproductive loss or of having an affected child. Genetic counselling before testing ensures that the individual is informed of the potential consequences of carrier testing including the option of prenatal diagnosis. In the presence of a family history, carrier testing is usually offered in the mid-teens when young people can decide whether they want to know their carrier status. For autosomal recessive conditions such as cystic fibrosis, some people may wish to wait until they have a partner so that testing can be done together, as there will be reproductive consequences only if both are found to be carriers.

Prenatal testing

The availability of prenatal genetic testing has enabled many couples at high genetic risk to embark upon pregnancies that they would otherwise have not undertaken. However, prenatal testing, and the associated option of termination of pregnancy, can have important psychological sequelae for pregnant women and their partners. In the presence of a known family history, genetic counselling is ideally offered in advance of pregnancy so that couples have time to make a considered choice. This also enables the laboratory to complete any family testing necessary before a prenatal test can be undertaken. Counselling should be provided within the antenatal setting when prenatal genetic tests are offered to couples without a previous family history, such as amniocentesis testing after a raised Down syndrome biochemical screening result. To help couples make an informed choice, information should be presented about the condition, the chance of it occurring, the test procedure and associated risks, the accuracy of the test, and the potential outcomes of testing including the option of termination of pregnancy. Couples at high genetic risk often require ongoing counselling and support during pregnancy. Psychologically, many couples cope with the uncertainty by remaining tentative about the pregnancy until receiving the test result. If the outcome of testing leads to termination of a wanted pregnancy, follow-up support should be offered. Even if favourable results are given, couples may still have some anxiety until the baby is born and clinical examination in the newborn period gives reassurance about normality. Occasionally, confirmatory investigations may be indicated.

Legal and ethical issues

There are many highly publicised controversies in genetics, including the use of modern genetic technologies in genetic testing, embryo research, gene therapy and the potential application of cloning techniques. In everyday clinical practice, however, the legal and ethical issues faced by professionals working in clinical genetics are generally similar to those in other specialities. Certain dilemmas are more specific to clinical genetics, for example, the issue of whether or not genetic information belongs to the individual and/or to other relatives remains controversial. Public perception of genetics is made more sensitive by past abuses, often carried out in the name of scientific progress. Whilst professionals have learnt lessons from history, the public may still have anxieties about the purpose of genetic services.

There is an extensive regulatory and advisory framework for biotechnology in the UK. The bodies that have particular

---

Decision by patient to proceed with test

Test session
- Written consent (including disclosure to GP)
- Clear arrangements for result giving and follow up

Period of 2–6 weeks

Result session
- Results communicated
- Arrangements for follow up confirmed
- Prompt written confirmation of result to patient and GP (if consent given)

timing agreed with patient

Follow-up
Mutation positive result
- Input from genetic service and primary care
- Planned medical surveillance anticipating future onset of symptoms
- Psychological support
- Genetic counselling for children at appropriate age

Mutation negative result
- Psychological support often needed towards adjustment and impact on relatives still at risk.

Figure 3.4 Protocol for presymptomatic testing for late onset disorders

---

Box 3.5

Prenatal detection of unexpected abnormalities
- serum biochemical screening
- routine ultrasonography
- amniocentesis for chromosomal analysis following abnormalities on biochemical or ultrasound screening

Prenatal diagnosis of abnormalities anticipated prior to pregnancy
- ultrasonography for known risk of specific congenital abnormality because of a previous affected child
- chromosomal analysis because of familial chromosome translocation or previous affected child
- molecular testing because of family history of single gene disorder

---

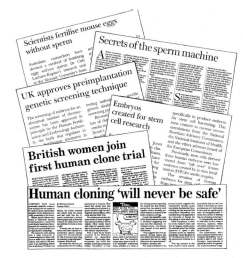

Figure 3.5 Advances in genetic technology generate debate about ethical issues

responsibility for clinical genetics are the Human Genetics Commission, the Gene Therapy Advisory Committee and the Genetics and the Insurance Committee.

Informed consent

Competent adults can give informed consent for a procedure when they have been given appropriate information by professionals and have had the chance to think about it. With regard to genetic tests, the information given needs to include the reason for the test (diagnostic or predictive), its accuracy and the implications of the result. It may be difficult to ensure that consent is truly informed when the patient is a child, or other vulnerable person, such as an individual with cognitive impairment. This is of most concern if the proposed genetic test is being carried out for the benefit of other members of the family who wish to have a genetic disorder confirmed in order to have their own risk assessed.

Genetic tests in childhood

In the UK the professional consensus is that a predictive genetic test should be carried out in childhood only when it is in the best interests of the child concerned. It is important to note that both medical and non-medical issues need to be considered when the child's best interests are being assessed. There may be a potential for conflict between the parents' "need to know" and the child's right to make his or her own decisions on reaching adulthood. In most cases, genetic counselling helps to resolve such situations without predictive genetic testing being carried out during childhood, since genetic tests for carrier state in autosomal recessive disorders only become of consequence at reproductive age, and physical examination to exclude the presence of clinical signs usually avoids the need for predictive genetic testing for late-onset dominant disorders.

Confidentiality

Confidentiality is not an absolute right. It may be breached, for example, if there is a risk of serious harm to others. In practice, however, it can be difficult to assess what constitutes serious harm. There is the potential for conflict between an individual's right to privacy and his or her genetic relatives' right to know information of relevance to themselves. Occasionally patients are reluctant to disclose a genetic diagnosis to other family members. In practice the individual's sense of responsibility to his or her relatives means that, in time, important information is shared within most families. There may also be conflict between an individual's right to privacy and the interests of other third parties, for example employers and insurance companies.

Unsolicited information

Problems may arise where unsolicited information becomes available. Non-paternity may be revealed either as a result of a genetic test, or through discussion with another family member. Where this would change the individual's genetic risk, the professional needs to consider whether to divulge this information and to whom. In other situations a genetic test, such as chromosomal analysis of an amniocentesis sample for Down syndrome, may reveal an abnormality other than the one being tested for. If this possibility is known before testing, it should be explained to the person being tested.

Non-directiveness

Non-directiveness is taken to be a cornerstone of contemporary genetic counselling practice, and is important in promoting

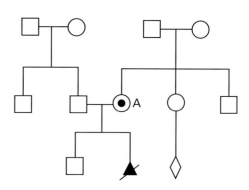

Fig. 3.6 Individual A was found to carry a balanced chromosomal rearrangement following termination of pregnancy for fetal abnormality. Initially she refused to inform her sister and brother about the potential risks to their future children, but decided to share this information when her sister became pregnant, so that she could have the opportunity to ask for tests

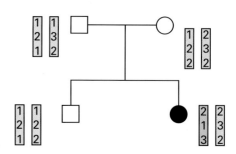

Figure 3.7 Genotyping using DNA markers linked to the SMN gene, undertaken to enable future prenatal diagnosis of spinal muscular atrophy, demonstrated that the affected child had inherited genetic markers not present in her father or mother, indicating non-paternity and affecting the risk of recurrence for future pregnancies.

autonomy of the individual. It is important for professionals to be aware that it may be difficult to be non-directive in certain situations, particularly where individuals or couples ask directly for advice. In general, genetic counsellors refrain from directing patients who are making reproductive or predictive test decisions, but there is an ongoing debate about whether it is possible for a professional to be non-directive, and whether such an approach is always appropriate for all types of decisions that need to be made by people with a family history of genetic disease.

4 Chromosomal analysis

The correct chromosome complement in humans was established in 1956, and the first chromosomal disorders (Down, Turner, and Klinefelter syndromes) were defined in 1959. Since then, refinements in techniques of preparing and examining samples have led to the description of hundreds of disorders that are due to chromosomal abnormalities.

Cell division

Most human somatic cells are diploid (2n=46), contain two copies of the genome and divide by mitosis. Germline oocytes and spermatocytes divide by meiosis to produce haploid gametes (n=23). Some human somatic cells, for example giant megakaryocytes, are polyploid and others, for example muscle cells, contain multiple diploid nuclei as a result of cell fusion.

During cell division the DNA of the chromosomes becomes highly condensed and they become visible under the light microscope as structures containing two chromatids joined together by a single centromere. This structure is essential for segregation of the chromosomes during cell division and chromosomes without centromeres are lost from the cell.

Chromosomes replicate themselves during the cell cycle which consists of a short M phase during which mitosis occurs, and a longer interphase. During interphase there is a G1 gap phase, an S phase when DNA synthesis occurs and a G2 gap phase. The stages of mitosis – prophase, prometaphase, metaphase, anaphase and telophase – are followed by cytokinesis when the cytoplasm divides to give two daughter cells. The process of mitosis produces two identical diploid daughter cells. Meiosis is also preceded by a single round of DNA synthesis, but this is followed by two cell divisions to produce the haploid gametes. The first division involves the pairing and separation of maternal and paternal chromosome homologs during which exchange of chromosomal material takes place. This process of recombination separates groups of genes that were originally located on the same chromosome and gives rise to individual genetic variation. The second cell division is the same as in mitosis, but there are only 23 chromosomes at the start of division. During spermatogenesis, each spermatocyte produces four spermatozoa, but during oogenesis there is unequal division of the cytoplasm, giving rise to the first and second polar bodies with the production of only one large mature egg cell.

Figure 4.1 Normal male chromosome constitution with idiograms demonstrating G banding pattern of each individual chromosome (courtesy of Dr Lorraine Gaunt and Helena Elliott, Regional Genetic Service, St Mary's Hospital, Manchester)

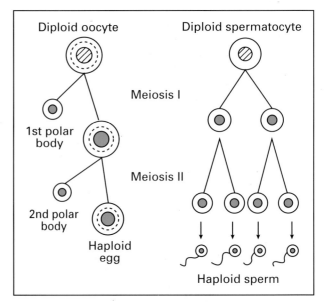

Figure 4.2 The process of meiosis in production of mature egg and sperm

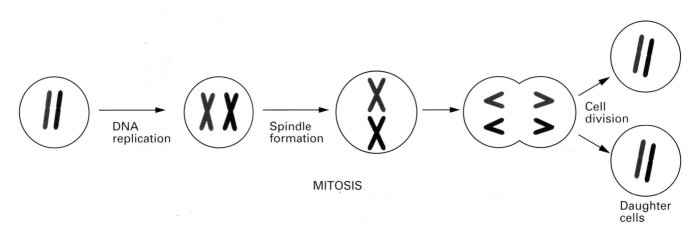

MITOSIS

Figure 4.3 Mitosis

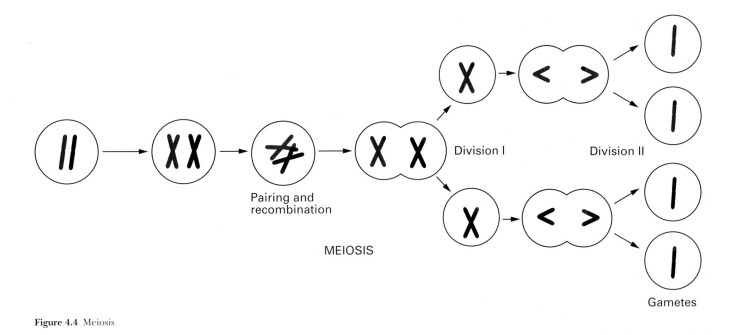

Figure 4.4 Meiosis

Chromosomal analysis

Chromosomal analysis is usually performed on white blood cell cultures. Other samples analysed on a routine basis include cultures of fibroblasts from skin biopsy samples, chorionic villi and amniocytes for prenatal diagnosis, and actively dividing bone marrow cells. The cell cultures are treated to arrest growth during metaphase or prometaphase when the chromosomes are visible. Until the 1970s, chromosomes could only be analysed on the basis of size and number. A variety of banding techniques are now possible and allow more precise identification of chromosomal rearrangements. The most commonly used is G-banding, in which the chromosomes are subjected to controlled trypsin digestion and stained with Giemsa to produce a specific pattern of light and dark bands for each chromosome.

The chromosome constitution of a cell is referred to as its karyotype and there is an International System for Human Cytogenetic Nomenclature (ISCN) for describing abnormalities. The Paris convention in 1971 defined the terminology used in reporting karyotypes. The centromere is designated "cen" and the telomere (terminal structure of the chromosome) as "ter". The short arm of each chromosome is designated "p" (petit) and the long arm "q" (queue). Each arm is subdivided into a number of bands and sub-bands depending on the resolution of the banding pattern achieved. High resolution cytogenetic techniques have permitted identification of small interstitial chromosome deletions in recognised disorders of previously unknown origin, such as Prader–Willi and Angelman syndromes. Deletions too small to be detected by microscopy may be amenable to diagnosis by molecular in situ hybridisation techniques.

Karyotypes are reported in a standard format giving the total number of chromosomes first, followed by the sex chromosome constitution. All cell lines are described in mosaic abnormalities, indicating the frequency of each. Additional or missing chromosomes are indicated by + or − for whole chromosomes, with an indication of the type of abnormality if there is a ring or marker chromosome. Structural rearrangements are described by indicating the p or q arm and the band position of the break points.

Table 4.1 Definitions

Euploid	Chromosome numbers are multiples of the haploid set (2n)
Polyploid	Chromosome numbers are greater than diploid (3n, triploid)
Aneuploid	Chromosome numbers are not exact multiples of the haploid set (2n+1 trisomy; 2n−1 monosomy)
Mosaic	Presence of two different cell lines derived from one zygote (46XX/45X, Turner mosaic)
Chimaera	Presence of two different cell lines derived from fusion of two zygotes (46XX/46XY, true hermaphrodite)

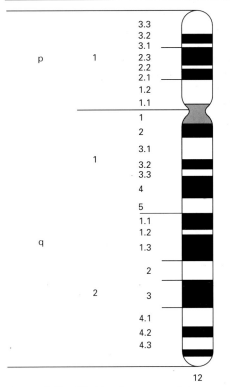

Figure 4.5 Simplified banding pattern of chromosome 12

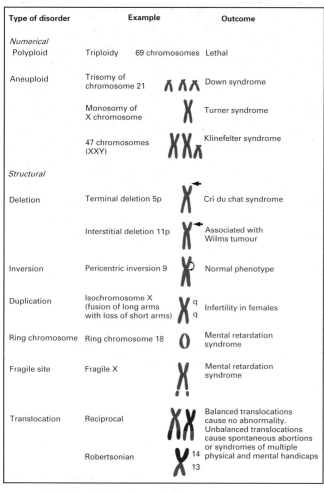

Type of disorder	Example	Outcome
Numerical		
Polyploid	Triploidy 69 chromosomes	Lethal
Aneuploid	Trisomy of chromosome 21	Down syndrome
	Monosomy of X chromosome	Turner syndrome
	47 chromosomes (XXY)	Klinefelter syndrome
Structural		
Deletion	Terminal deletion 5p	Cri du chat syndrome
	Interstitial deletion 11p	Associated with Wilms tumour
Inversion	Pericentric inversion 9	Normal phenotype
Duplication	Isochromosome X (fusion of long arms with loss of short arms)	Infertility in females
Ring chromosome	Ring chromosome 18	Mental retardation syndrome
Fragile site	Fragile X	Mental retardation syndrome
Translocation	Reciprocal	Balanced translocations cause no abnormality. Unbalanced translocations cause spontaneous abortions or syndromes of multiple physical and mental handicaps
	Robertsonian	

Figure 4.7 Types of chromosomal abnormality

Table 4.2 Reporting of karyotypes

Total number of chromosomes given first followed by sex chromosome constitution

46,XX	Normal female
47,XXY	Male with Klinefelter syndrome
47,XXX	Female with triple X syndrome

Additional or lost chromosomes indicated by + or −

47,XY,+21	Male with trisomy 21 (Down syndrome)
46,XX,12p+	Additional unidentified material on short arm of chromosome 12

All cell lines present are shown for mosaics

46,XX/47,XX,+21	Down syndrome mosaic
46,XX/47,XXX/45,X	Turner/triple X syndrome mosaic

Structural rearrangements are described, identifying p and q arms and location of abnormality

46,XY,del11(p13)	Deletion of short arm of chromosome 11 at band 13
46,XX,t(X;7)(p21;q23)	Translocation between chromosomes X and 7 with break points in respective chromosomes

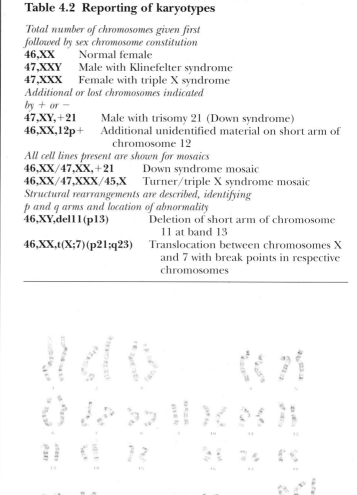

Figure 4.6 47, XXX karyotype in triple X syndrome (courtesy of Dr Lorraine Gaunt and Helena Elliott, Regional Genetic Service, St Mary's Hospital, Manchester)

Molecular cytogenetics

Fluorescence in situ hybridisation (FISH) is a recently developed molecular cytogenetic technique, involving hybridisation of a DNA probe to a metaphase chromosome spread. Single stranded probe DNA is fluorescently labelled using biotin and avidin and hybridised to the denatured DNA of intact chromosomes on a microscope slide. The resultant DNA binding can be seen directly using a fluorescence microscope.

One application of FISH is in chromosome painting. This technique uses an array of specific DNA probes derived from a whole chromosome and causes the entire chromosome to fluoresce. This can be used to identify the chromosomal origin of structural rearrangements that cannot be defined by conventional cytogenetic techniques.

Alternatively, a single DNA probe corresponding to a specific locus can be used. Hybridisation reveals fluorescent spots on each chromatid of the relative chromosome. This method is used to detect the presence or absence of specific DNA sequences and is useful in the diagnosis of syndromes caused by sub-microscopic deletions, such as William syndrome, or in identifying carriers of single gene defects due to large deletions, such as Duchenne muscular dystrophy.

It is possible to use several separate DNA probes, each labelled with a different fluorochrome, to analyse more than

Figure 4.8 Fluorescence in situ hybridisation of normal metaphase chromosomes hybridised with chromosome 20 probes derived from the whole chromosome, which identify each individual chromosome 20 (courtesy of Dr Lorraine Gaunt, Regional Genetic Service, St Mary's Hospital, Manchester)

one locus or chromosome region in the same reaction. Another application of this technique is in the study of interphase nuclei, which permits the study of non-dividing cells. Thus, rapid results can be obtained for the diagnosis or exclusion of Down syndrome in uncultured amniotic fluid samples using chromosome 21 specific probes.

Incidence of chromosomal abnormalities

Chromosomal abnormalities are particularly common in spontaneous abortions. At least 20% of all conceptions are estimated to be lost spontaneously, and about half of these are associated with a chromosomal abnormality, mainly autosomal trisomy. Cytogenetic studies of gametes have shown that 10% of spermatozoa and 25% of mature oocytes are chromosomally abnormal. Between 1 and 3% of all recognised conceptions are triploid. The extra haploid set is usually due to fertilisation of a single egg by two separate sperm. Very few triploid pregnancies continue to term and postnatal survival is not possible unless there is mosaicism with a normal cell line present as well. All autosomal monosomies and most autosomal trisomies are also lethal in early embryonic life. Trisomy 16, for example, is frequently detected in spontaneous first trimester abortuses, but never found in liveborn infants.

In liveborn infants chromosomal abnormalities occur in about 9 per 1000 births. The incidence of unbalanced abnormalities affecting autosomes and sex chromosomes is about the same. The effect on the child depends on the type of abnormality. Balanced rearrangements usually have no phenotypic effect. Aneuploidy affecting the sex chromosomes is fairly frequent and the effect much less severe than in autosomal abnormalities. Unbalanced autosomal abnormalities cause disorders with multiple congenital malformations, almost invariably associated with mental retardation.

Figure 4.9 Fluorescence in situ hybridisation of metaphase chromosomes from a male with 46 XX chromosome constitution hybridised with separate probes derived from both X and Y chromosomes. The X chromosome probe (yellow) has hybridised to both X chromosomes. The Y chromosome probe (red) has hybridised to one of the X chromosomes, which indicates that this chromosome carries Y chromosomal DNA, thus accounting for the subject's phenotypic sex (courtesy of Dr Lorraine Gaunt, Regional Genetic Service, St Mary's Hospital, Manchester)

Table 4.3 Frequency of chromosomal abnormalities in spontaneous abortions and stillbirths (%)

Spontaneous abortions	
All	50
Before 12 weeks	60
12–20 weeks	20
Stillbirths	5

Table 4.4 Frequency of chromosomal abnormalities in newborn infants (%)

All	0.91
Autosomal trisomy	0.14
Autosomal rearrangements	
balanced	0.52
unbalanced	0.06
Sex chromosome abnormality	0.19

5 Common chromosomal disorders

Abnormalities of the autosomal chromosomes generally cause multiple congenital malformations and mental retardation. Children with more than one physical abnormality and developmental delay or learning disability should therefore undergo chromosomal analysis as part of their investigation. Chromosomal disorders are incurable but most can be reliably detected by prenatal diagnostic techniques. Amniocentesis or chorionic villus sampling should be offered to women whose pregnancies are at increased risk – namely, couples in whom one partner carries a balanced translocation, women identified by biochemical screening for Down syndrome and couples who already have an affected child. Unfortunately, when there is no history of previous abnormality the risk in many affected pregnancies cannot be predicted before the child is born.

Down syndrome

Down syndrome, due to trisomy 21, is the commonest autosomal trisomy with an overall incidence in liveborn infants of between 1 in 650 and 1 in 800. Most conceptions with trisomy 21 do not survive to term. Two thirds of conceptions miscarry by mid-trimester and one third of the remainder subsequently die in utero before term. The survival rate for liveborn infants is surprisingly high with 85% surviving into their 50s. Congenital heart defects remain the major cause of early mortality, but additional factors include other congenital malformations, respiratory infections and the increased risk of leukaemia.

An increased risk of Down syndrome may be identified prenatally by serum biochemical screening tests or by detection of abnormalities by ultrasound scanning. Features indicating an increased risk of Down syndrome include increased first trimester nuchal translucency or thickening, structural heart defects and duodenal atresia. Less specific features include choroid plexus cysts, short femori and humeri, and echogenic bowel. In combination with other risk factors their presence indicates the need for diagnostic prenatal chromosome tests.

The facial appearance at birth usually suggests the presence of the underlying chromosomal abnormality, but clinical diagnosis can be difficult, especially in premature babies, and should always be confirmed by cytogenetic analysis. In addition to the facial features, affected infants have brachycephaly, a short neck, single palmar creases, clinodactyly, wide gap between the first and second toes, and hypotonia. Older children are often described as being placid, affectionate and music-loving, but they display a wide range of behavioural and personality traits. Developmental delay becomes more apparent in the second year of life and most children have moderate learning disability, although the IQ can range from 20 to 85. Short stature is usual in older children and hearing loss and visual problems are common. The incidence of atlanto-axial instability, hypothyroidism and epilepsy is increased. After the age of 40 years, neuropathological changes of Alzheimer disease are almost invariable.

Down syndrome risk

Most cases of Down syndrome (90%) are due to nondisjunction of chromosome 21 arising during the first meiotic cell division in oogenesis. A small number of cases arise in meiosis II during oogenesis, during spermatogenesis or during mitotic cell division in the early zygote. Although occurring at any age, the

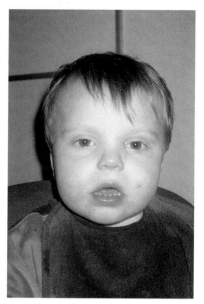

Figure 5.1 Child with dysmorphic facial features and developmental delay due to deletion of chromosome 18(18q-)

Figure 5.2 Trisomy 21 (47, XX + 21) in Down syndrome (courtesy of Dr Lorraine Gaunt and Helena Elliott, Regional Genetic Service, St Mary's Hospital, Manchester)

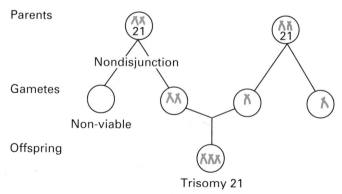
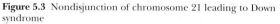

Figure 5.3 Nondisjunction of chromosome 21 leading to Down syndrome

Box 5.1 Risk of Down syndrome in livebirths and at amniocentesis

Maternal age (at delivery or amniocentesis)	Liveborn risk	Risk at amniocentesis
All ages	1 in 650	
Age 30	1 in 900	
Age 35	1 in 385	1 in 256
Age 36	1 in 305	1 in 200
Age 37	1 in 240	1 in 156
Age 38	1 in 190	1 in 123
Age 39	1 in 145	1 in 96
Age 40	1 in 110	1 in 75
Age 44	1 in 37	1 in 29

risk of having a child with trisomy 21 increases with maternal age. This age-related risk has been recognised for a long time, but the underlying mechanism is not understood. The risk of recurrence for any chromosomal abnormality in a liveborn infant after the birth of a child with trisomy 21 is increased by about 1% above the population age related risk. The risk is probably 0.5% for trisomy 21 and 0.5% for other chromosomal abnormalities. This increase in risk is more significant for younger women. In women over the age of 35 the increase in risk related to the population age-related risk is less apparent. Population risk tables for Down syndrome and other trisomies have been derived from the incidence in livebirths and the detection rate at amniocentesis. Because of the natural loss of affected pregnancies, the risk for livebirths is less than the risk at the time of prenatal diagnosis. Although the majority of males with Down syndrome are infertile, affected females who become pregnant have a high risk (30–50%) of having a Down syndrome child.

Translocation Down syndrome

About 5% of cases of Down syndrome are due to translocation, in which chromosome 21 is translocated onto chromosome 14 or, occasionally, chromosome 22. In less than half of these cases one of the parents has a balanced version of the same translocation. A healthy adult with a balanced translocation has 45 chromosomes, and the affected child has 46 chromosomes, the extra chromosome 21 being present in the translocation form. The risk of Down syndrome in offspring is about 10% when the balanced translocation is carried by the mother and 2.5% when carried by the father. If neither parent has a balanced translocation, the chromosomal abnormality in an affected child represents a spontaneous, newly arising event, and the risk of recurrence is low (<1%). Recurrence due to parental gonadal mosaicism cannot be completely excluded.

Occasionally, Down syndrome is due to a 21;21 translocation. Some of these cases are due to the formation of an isochromosome following the fusion of sister chromatids. In cases of true 21;21 Robertsonian translocation, a parent who carries the balanced translocation would be unable to have normal children (see figure 5.6).

When a case of translocation Down syndrome occurs it is important to test other family members to identify all carriers of the translocation whose pregnancies would be at risk. Couples concerned about a family history of Down syndrome can have their chromosomes analysed from a sample of blood to exclude a balanced translocation if the karyotype of the affected person is not known. This usually avoids unnecessary amniocentesis during pregnancy.

Figure 5.4 Down syndrome due to Robertsonian translocation between chromosomes 14 and 21 (courtesy of Dr Lorraine Gaunt and Helena Elliott, Regional Genetic Service, St Mary's Hospital, Manchester)

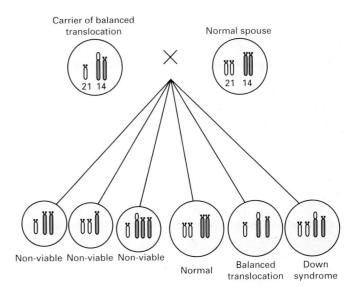

Figure 5.5 Possibilities for offspring of a 14;21 Robertsonian translocation carrier

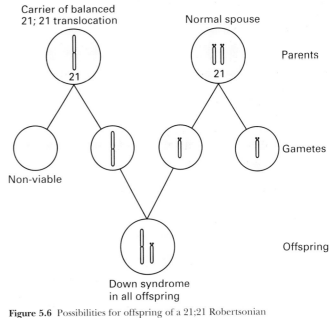

Figure 5.6 Possibilities for offspring of a 21;21 Robertsonian translocation carrier

19

Other autosomal trisomies

Trisomy 18 (Edwards syndrome)

Trisomy 18 has an overall incidence of around 1 in 6000 live births. As with Down syndrome most cases are due to nondisjunction and the incidence increases with maternal age. The majority of trisomy 18 conceptions are lost spontaneously with only about 2.5% surviving to term. Many cases are now detectable by prenatal ultasound scanning because of a combination of intrauterine growth retardation, oligohydramnios or polyhydramnios and major malformations that indicate the need for amniocentesis. About one third of cases detected during the second trimester might survive to term. The main features of trisomy 18 include growth deficiency, characteristic facial appearance, clenched hands with overlapping digits, rocker bottom feet, cardiac defects, renal abnormalities, exomphalos, myelomeningocele, oesophageal atresia and radial defects. Ninety percent of affected infants die before the age of 6 months but 5% survive beyond the first year of life. All survivors have severe mental and physical disability. The risk of recurrence for any trisomy is probably about 1% above the population age-related risk. Recurrence risk is higher in cases due to a translocation where one of the parents is a carrier.

Trisomy 13 (Patau syndrome)

The incidence of trisomy 13 is about 1 per 15 000 live births. The majority of trisomy 13 conceptions spontaneously abort in early pregnancy. About 75% of cases are due to nondisjunction, and are associated with a similar overall risk for recurrent trisomy as in trisomy 18 and 21 cases. The remainder are translocation cases, usually involving 13;14 Robertsonian translocations. Of these, half arise de novo and half are inherited from a carrrier parent. The frequency of 13;14 translocations in the general population is around 1 in 1000 and the risk of a trisomic conception for a carrier parent appears to be around 1%. The risk of recurrence after the birth of an affected child is low but difficult to determine. Prenatal ultrasound scanning will detect abnormalities leading to a diagnosis in about 50% of cases. Most liveborn affected infants succumb within hours or weeks of delivery. Eighty percent die within 1 month, 3% survive to 6 months. The main features of trisomy 13 include structural abnormalities of the brain, particularly microcephaly and holoprosencephaly (a developmental defect of the forebrain), facial and eye abnormalities, cleft lip and palate, postaxial polydactyly, congenital heart defects, renal abnormalities, exomphalos and scalp defects. Survivors have very severe mental and physical disability, usually with associated epilepsy, blindness and deafness.

Chromosomal mosaicism

After fertilisation of a normal egg nondisjunction may occur during a mitotic division in the developing embryo giving rise to daughter cells that are trisomic and nulisomic for the chromosome involved in the disjunction error. The nulisomic cell would not be viable, but further cell division of the trisomic cell, along with those of the normal cells, leads to chromosomal mosaicism in the fetus. Alternatively a chromosome may be lost from a cell in an embryo that was trisomic for that chromosome at conception. Further division of this cell would lead to a population of cells with a normal karyotype, again resulting in mosaicism. In Down syndrome mosaicism, for example, one cell line has a normal constitution of 46 chromosomes and the other has a constitution of 47 + 21.

Figure 5.7 Trisomy 18 mosaicism can be associated with mild to moderate developmental delay without congenital malformations or obvious dysmorphic features

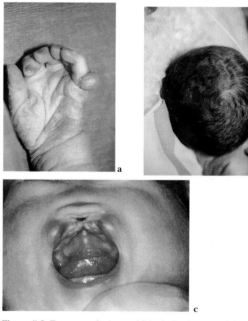

Figure 5.8 Features of trisomy 13 include a) post-axial polydactyly b)scalp defects and c)mid-line cleft lip and palate (courtesy of Professor Dian Donnai, Regional Genetic Service, St Mary's Hospital, Manchester)

Figure 5.9 Mild facial dysmorphism in a girl with mosaic trisomy 21

The proportion of each cell line varies among different tissues and this influences the phenotypic expression of the disorder. The severity of mosaic disorders is usually less than non-mosaic cases, but can vary from virtually normal to a phenotype indistinguishable from full trisomy. In subjects with mosaic chromosomal abnormalities the abnormal cell line may not be present in peripheral lymphocytes. In these cases, examination of cultured fibroblasts from a skin biopsy specimen is needed to confirm the diagnosis.

The clinical effect of a mosaic abnormality detected prenatally is difficult to predict. Most cases of mosaicism for chromosome 20 detected at amniocentesis, for example, are not associated with fetal abnormality. The trisomic cell line is often confined to extra fetal tissues, with neonatal blood and fibroblast cultures revealing normal karyotypes in infants subsequently delivered at term. In some cases, however, a trisomic cell line is detected in the infant after birth and this may be associated with physical abnormalities or developmental delay.

Mosaicism for a marker (small unidentified) chromosome carries a much smaller risk of causing mental retardation if familial, and therefore the parents need to be investigated before advice can be given. Chromosomal mosaicism detected in chorionic villus samples often reflects an abnormality confined to placental tissue that does not affect the fetus. Further analysis with amniocentesis or fetal blood sampling may be indicated together with detailed ultrasound scanning.

Translocations

Robertsonian translocations

Robertsonian translocations occur when two of the acrocentric chromosomes (13, 14, 15, 21, or 22) become joined together. Balanced translocation carriers have 45 chromosomes but no significant loss of overall chromosomal material and they are almost always healthy. In unbalanced translocation karyotypes there are 46 chromosomes with trisomy for one of the chromosomes involved in the translocation. This may lead to spontaneous miscarriage (chromosomes 14, 15, and 22) or liveborn infants with trisomy (chromosomes 13 and 21). Unbalanced Robertsonian translocations may arise spontaneously or be inherited from a parent carrying a balanced translocation. (Translocation Down syndrome is discussed earlier in this chapter.)

Reciprocal translocations

Reciprocal translocations involve exchange of chromosomal segments between two different chromosomes, generated by the chromosomes breaking and rejoining incorrectly. Balanced reciprocal translocations are found in one in 500–1000 healthy people in the population. When an apparently balanced recriprocal translocation is detected at amniocentesis it is important to test the parents to see whether one of them carries the same translocation. If one parent is a carrier, the translocation in the fetus is unlikely to have any phenotypic effect. The situation is less certain if neither parent carries the translocation, since there is some risk of mental disability or physical effect associated with de novo translocations from loss or damage to the DNA that cannot be seen on chromosomal analysis. If the translocation disrupts an autosomal dominant or X linked gene, it may result in a specific disease phenotype.

Once a translocation has been identified it is important to investigate relatives of that person to identify other carriers of

Figure 5.10 Normal 8 month old infant born after trisomy 20 mosaicism detected in amniotic cells. Neonatal blood sample showed normal karyotype

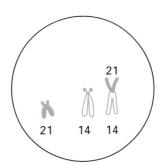 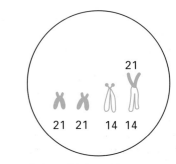

Figure 5.11 Balanced Robertsonian translocation affecting chromosomes 14 and 21

Unbalanced Robertsonian translocation affecting chromosomes 14 and 21 and resulting in Down syndrome

Figure 5.12 Balanced Robertsonian translocation affecting chromosomes 13 and 14 (courtesy of Dr Lorraine Gaunt and Helena Elliott, Regional Genetic Service, St Mary's Hospital, Manchester)

the balanced translocation whose offspring would be at risk. Abnormalities resulting from an unbalanced reciprocal translocation depend on the particular chromosomal fragments that are present in monosomic or trisomic form. Sometimes spontaneous abortion is inevitable; at other times a child with multiple abnormalities may be born alive. Clinical syndromes have been described due to imbalance of some specific chromosomal segments. This applies particularly to terminal chromosomal deletions. For other rearrangements, the likely effect can only be assessed from reports of similar cases in the literature. Prediction is never precise, since reciprocal translocations in unrelated individuals are unlikely to be identical at the molecular level and other factors may influence expression of the chromosomal imbalance. The risk of an unbalanced karyotype occurring in offspring depends on the individual translocation and can also be difficult to determine. An overall risk of 5–10% is often quoted. After the birth of one affected child, the recurrence risk is generally higher (5–30%). The risk of a liveborn affected child is less for families ascertained through a history of recurrent pregnancy loss where there have been no liveborn affected infants. Pregnancies at risk can be monitored with chorionic villus sampling or amniocentesis.

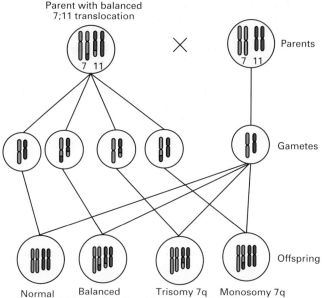

Figure 5.13 Possibilities for offspring of a 7;11 reciprocal translocation carrier

Deletions

Chromosomal deletions may arise de novo as well as resulting from unbalanced translocations. De novo deletions may affect the terminal part of the chromosome or an interstitial region. Recognisable syndromes have been delineated for the most commonly occurring deletions. The best known of these are cri du chat syndrome caused by a terminal deletion of the short arm of chromosome 5 (5p-) and Wolf–Hirschhorn syndrome caused by a terminal deletion of the short arm of chromosome 4 (4p-).

Microdeletions
Several genetic syndromes have now been identified by molecular cytogenetic techniques as being due to chromosomal deletions too small to be seen by conventional analysis. These are termed submicroscopic deletions or microdeletions and probably affect less than 4000 kilobases of DNA. A microdeletion may involve a single gene, or extend over several genes. The term contiguous gene syndrome is applied when several genes are affected, and in these disorders the features present may be determined by the extent of the deletion. The chromosomal location of a microdeletion may be initially identified by the presence of a larger visible cytogenetic deletion in a proportion of cases, as in Prader–Willi and Angelman syndrome, or by finding a chromosomal translocation in an affected individual, as occured in William syndrome.

A microdeletion on chromosome 22q11 has been found in most cases of DiGeorge syndrome and velocardiofacial syndrome, and is also associated with certain types of isolated congenital heart disease. With an incidence of 8 per 1000 live births, congenital heart disease is one of the most common birth defects. The aetiology is usually unknown and it is therefore important to identify cases caused by 22q11 deletion. Isolated cardiac defects due to microdeletions of chromosome 22q11 often include outflow tract abnormalities. Deletions have been observed in both sporadic and familial cases and are responsible for about 30% of non-syndromic conotruncal malformations including interrupted aortic arch, truncus arteriosus and tetralogy of Fallot.

Figure 5.14 Cri du chat syndrome associated with deletion of short arm of chromosome 5 (courtesy of Dr Lorraine Gaunt and Helena Elliott, Regional Genetic Service, St Mary's Hospital, Manchester)

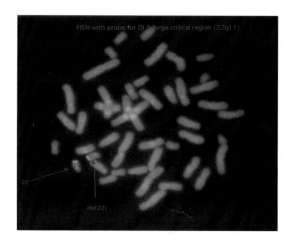

Figure 5.15 Fluorescence in situ hybridisation with a probe from the DiGeorge critical region of chromosome 22q11, which shows hybridisation to only one chromosome 22 (red signal), thus indicating that the other chromosome 22 is deleted in this region (courtesy of Dr Lorraine Gaunt and Helen Elliott, Regional Genetic Service, St Mary's Hospital, Manchester)

DiGeorge syndrome involves thymic aplasia, parathyroid hypoplasia, aortic arch and conotruncal anomalies, and characteristic facies due to defects of 3rd and 4th branchial arch development. Velocardiofacial syndrome was described as a separate clinical entity, but does share many features in common with DiGeorge syndrome. The features include mild mental retardation, short stature, cleft palate or speech defect from palatal dysfunction, prominent nose and congenital cardiac defects including ventricular septal defect, right sided aortic arch and tetralogy of Fallot.

Sex chromosome abnormalities

Numerical abnormalities of the sex chromosomes are fairly common and their effects are much less severe than those caused by autosomal abnormalities. Sex chromosome abnormalites are often detected coincidentally at amniocentesis or during investigation for infertility. Many cases are thought to cause no associated problems and to remain undiagnosed. The risk of recurrence after the birth of an affected child is very low. When more than one additional sex chromosome is present learning disability or physical abnormality is more likely.

Turner syndrome

Turner syndrome is caused by the loss of one X chromosome (usually paternal) in fetal cells, producing a female conceptus with 45 chromosomes. This results in early spontaneous loss of the fetus in over 95% of cases. Severely affected fetuses who survive to the second trimester can be detected by ultrasonography, which shows cystic hygroma, chylothorax, asictes and hydrops. Fetal mortality is very high in these cases.

The incidence of Turner syndrome in liveborn female infants is 1 in 2500. Phenotypic abnormalities vary considerably but are usually mild. In some infants the only detectable abnormality is lymphoedema of the hands and feet. The most consistent features of the syndrome are short stature and infertility from streak gonads, but neck webbing, broad chest, cubitus valgus, coarctation of the aorta, renal anomalies and visual problems may also occur. Intelligence is usually within the normal range, but a few girls have educational or behavioural problems. Associations with autoimmune thyroiditis, hypertension, obesity and non-insulin dependent diabetes have been reported. Growth can be stimulated with androgens or growth hormone, and oestrogen replacement treatment is necessary for pubertal development. A proportion of girls with Turner syndrome have a mosaic 46XX/45X karyotype and some of these have normal gonadal development and are fertile, although they have an increased risk of early miscarriage and of premature ovarian failure. Other X chromosomal abnormalities including deletions or rearrangements can also result in Turner syndrome.

Triple X syndrome

The triple X syndrome occurs with an incidence of 1 in 1200 liveborn female infants and is often a coincidental finding. The additional chromosome usually arises by a nondisjunction error in maternal meiosis I. Apart from being taller than average, affected girls are physically normal. Educational problems are encountered more often in this group than in the other types of sex chromosome abnormalities. Mild delay with early motor and language development is fairly common and deficits in both receptive and expressive language persist into adolescence and adulthood. Mean intelligence quotient is often about 20 points lower than that in siblings and many girls require remedial teaching although the majority attend mainstream

Box 5.2 Examples of syndromes associated with microdeletions

Syndrome	Chromosomal deletion
DiGeorge	22q11
Velocardiofacial	22q11
Prader–Willi	15q11-13
Angelman	15q11-13
William	7q11
Miller–Dieker (lissencephaly)	17p13
WAGR (Wilms tumour + aniridia)	11p13
Rubinstein–Taybi	16p13
Alagille	20p12
Trichorhinophalangeal	8q24
Smith–Magenis	17p11

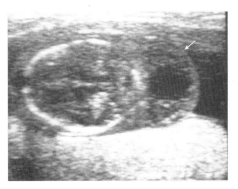

Figure 5.16 Cystic hygroma in Turner syndrome detected by ultrasonography (courtesy of Dr Sylvia Rimmer, Department of Radiology, St. Mary's Hospital, Manchester)

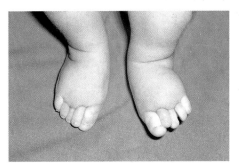

Figure 5.17 Lymphoedema of the feet may be the only manifestation of Turner syndrome in newborn infant

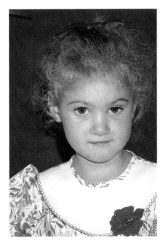

Figure 5.18 Normal appearance and development in 3-year old girl with triple X syndrome

schools. The incidence of mild psychological disturbances may be increased. Occasional menstrual problems are reported, but most triple X females are fertile and have normal offspring. Early menopause from premature ovarian failure may occur.

Klinefelter syndrome

The XXY karyotype of Klinefelter syndrome occurs with an incidence of 1 in 600 liveborn males. It arises by nondisjunction and the additional X chromosome is equally likely to be maternally or paternally derived. There is no increased early pregnancy loss associated with this karyotype. Many cases are never diagnosed. The primary feature of the syndrome is hypogonadism. Pubertal development usually starts spontaneously, but testicular size decreases from mid-puberty and hypogonadism develops. Testosterone replacement is usually required and affected males are infertile. Poor facial hair growth is an almost constant finding. Tall stature is usual and gynaecomastia may occur. The risk of cancer of the breast is increased compared to XY males. Intelligence is generally within the normal range but may be 10–15 points lower than siblings. Educational difficultes are fairly common and behavioural disturbances are likely to be associated with exposure to stressful environments. Shyness, immaturity and frustration tend to improve with testosterone replacement therapy.

XYY syndrome

The XYY syndrome occurs in about 1 per 1000 liveborn male infants, due to nondisjunction at paternal meiosis II. Fetal loss rate is very low. The majority of males with this karyotype have no evidence of clinical abnormality and remain undiagnosed. Accelerated growth in early childhood is common, leading to tall stature, but there are no other physical manifestations of the condition apart from the occasional reports of severe acne. Intelligence is usually within the normal range but may be about 10 points lower than in siblings and learning difficulties may require additional input at school. Behavioural problems can include hyperactivity, distractability and impulsiveness. Although initially found to be more prevalent among inmates of high security institutions, the syndrome is much less strongly associated with aggressive behaviour than previously thought although there is an increase in the risk of social maladjustment.

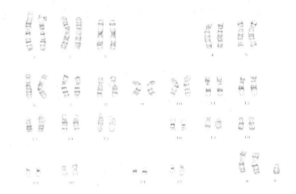

Figure 5.19 47, XXY karyotype in Klinefelter syndrome (courtesy of Dr Lorraine Gaunt and Helena Elliott, Regional Genetic Service, St. Mary's Hospital Manchester)

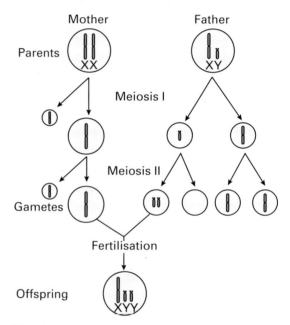

Figure 5.20 Nondisjunction error at paternal meiosis II resulting in XYY syndrome in offspring

6   Mendelian inheritance

Disorders caused by a defect in a single gene follow the patterns of inheritance described by Mendel and the term mendelian inheritance has been used to denote unifactorial inheritance since 1901. Individual disorders of this type are often rare, but are important because they are numerous. By 2001, over 9000 established gene or phenotype loci were listed in OMIM. Online Mendelian Inheritance in Man (TM). McKusick-Nathans Institute for Genetic Medicine, Johns Hopkins University (Baltimore, MD) and National Center for Biotechnology Information, National Library of Medicine (Bethesda, MD), 2000. World Wide Web URL: http://www.ncbi.nlm.nih.gov/omim. Risks within an affected family are often high and are calculated by knowing the mode of inheritance and the structure of the family pedigree.

Autosomal dominant inheritance

Autosomal dominant disorders affect both males and females. Mild or late onset conditions can often be traced through many generations of a family. Affected people are heterozygous for the abnormal allele and transmit this to half their offspring, whether male or female. The disorder is not transmitted by family members who are unaffected themselves. Estimation of risk is therefore apparently simple, but in practice several factors may cause difficulties in counselling families.

Late onset disorders

Dominant disorders may have a late or variable age of onset of signs and symptoms. People who inherit the defective gene will be destined to become affected, but may remain asymptomatic well into adult life. Young adults at risk may not know whether they have inherited the disorder and be at risk of transmitting it to their children at the time they are planning their own families. The possibility of detecting the mutant gene before symptoms become apparent has important consequences for conditions such as Huntington disease and myotonic dystrophy. Predictive genetic testing is considered in chapter 3.

Variable expressivity

The severity of many autosomal dominant conditions varies considerably between different affected individuals within the same family, a phenomenon referred to as variable expressivity. In some disorders this variability is due to instability of the underlying mutation, as in the disorders caused by trinucleotide repeat mutations (discussed in chapter 7). In many cases, the variability is unexplained. The likely severity in any affected individual is difficult to predict. A mildly affected parent may have a severely affected child, as illustrated by tuberous sclerosis, in which a parent with only skin manifestations of the disorder may have an affected child with infantile spasms and severe mental retardation. Tuberous sclerosis also demonstrates pleiotropy, resulting in a variety of apparently unrelated phenotypic features, such as skin hypopigmentation, multiple hamartomas and learning disability. Each of these pleiotropic effects can demonstrate variable expressivity and penetrance in a given family.

Penetrance

A few dominant disorders show lack of penetrance, that is, a person who inherits the gene does not develop the disorder. This phenomenon has been well documented in retinoblastoma,

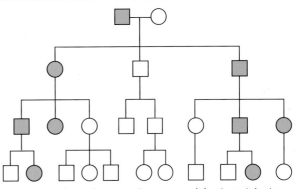

Figure 6.1 Pedigree demonstrating autosomal dominant inheritance

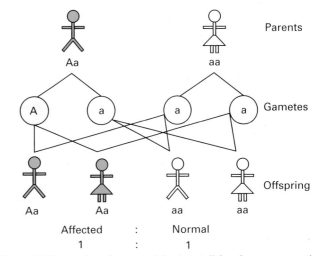

Figure 6.2 Segregation of autosomal dominant alleles when one parent is affected

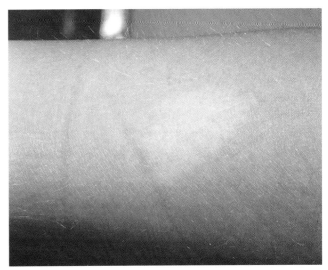

Figure 6.3 Ash leaf depigmentation may be the only sign of tuberous sclerosis in the parent of a severely affected child

otosclerosis and hereditary pancreatitis. In retinoblastoma, non-penetrance arises because a second somatic mutation needs to occur before a person who inherits the gene develops an eye tumour. For disorders that demonstrate non-penetrance, unaffected individuals cannot be completely reassured that they will not transmit the disorder to their children. This risk is fairly low (not exceeding 10%) because a clinically unaffected person is unlikely to be a carrier if the penetrance is high, and the chance of a gene carrier developing symptoms is small if the penetrance is low. Non-genetic factors may also influence the expression and penetrance of dominant genes, for example diet in hypercholesterolaemia, drugs in porphyria and anaesthetic agents in malignant hyperthermia.

New mutations

New mutations may account for the presence of a dominant disorder in a person who does not have a family history of the disease. New mutations are common in some disorders, such as achondroplasia, neurofibromatosis (NF1) and tuberous sclerosis, and rare in others, such as Huntington disease and myotonic dystrophy. When a disorder arises by new mutation, the risk of recurrence in future pregnancies for the parents of the affected child is very small. Care must be taken to exclude mild manifestations of the condition in one or other parent before giving this reassurance. This causes no problems in conditions such as achondroplasia that show little variability, but can be more difficult in many other conditions, such as neurofibromatosis and tuberous sclerosis. It is also possible that an apparently normal parent may carry a germline mutation. In some cases the mutation will be confined to gonadal tissue, with the parent being unaffected clinically. In others the mutation will be present in some somatic cells as well. In disorders with cutaneous manifestations, such as NF1, this may lead to segmental or patchy involvement of the skin. In either case, there will be a considerable risk of recurrence in future children. A dominant disorder in a person with a negative family history may alternatively indicate non-paternity.

Homozygosity

Homozygosity for dominant genes is uncommon, occurring only when two people with the same disorder have children together. This may happen preferentially with certain conditions, such as achondroplasia. Homozygous achondroplasia is a lethal condition and the risks to the offspring of two affected parents are 25% for being an affected homozygote (lethal), 50% for being an affected heterozygote, and 25% for being an unaffected homozygote. Thus two out of three surviving children will be affected.

Autosomal recessive inheritance

Most mutations inactivate genes and act recessively. Autosomal recessive disorders occur in individuals who are homozygous for a particular recessive gene mutation, inherited from healthy parents who carry the mutant gene in the heterozygous state. The risk of recurrence for future offspring of such parents is 25%. Unlike autosomal dominant disorders there is usually no preceding family history. Although the defective gene is passed from generation to generation, the disorder appears only within a single sibship, that is, within one group of brothers and sisters. The offspring of an affected person must inherit one copy of the mutant gene from them, but are unlikely to inherit a similar mutant gene from the other parent unless the gene is particularly prevalent in the population, or the parents

---

Box 6.2 Examples of autosomal dominant disorders

Achondroplasia
Acute intermittent porphyria
Charcot–Marie–Tooth disease
Facioscapulohumeral dystrophy
Familial adenomatous polyposis
Familial breast cancer (*BRCA 1, 2*)
Familial hypercholesterolaemia
Huntington disease
Myotonic dystrophy
Noonan syndrome
Neurofibromatosis (types 1 and 2)
Osteogenesis imperfecta
Spinocerebellar ataxia
Tuberous sclerosis

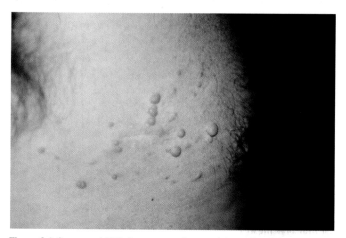

Figure 6.4 Segmental NF1 due to somatic mutation confined to one region of the body. This patient had no skin manifestations elsewhere

---

Box 6.3 Characteristics of autosomal dominant inheritance

- Males and females equally affected
- Disorder transmitted by both sexes
- Successive generations affected
- Male to male transmission occurs

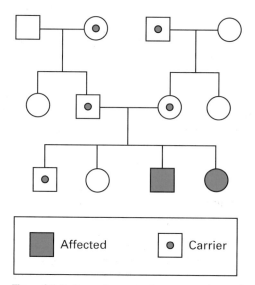

Affected		Carrier

Figure 6.5 Pedigree demonstrating autosomal recessive inheritance

are consanguineous. In most cases, therefore, the offspring of an affected person are not affected.

Autosomal recessive disorders are commonly severe, and many of the recognised inborn errors of metabolism follow this type of inheritance. Many complex malformation syndromes are also due to autosomal recessive gene mutations and their recognition is important in the first affected child in the family because of the high recurrence risk. Prenatal diagnosis for recessive disorders may be possible by performing biochemical assays, DNA analysis, or looking for structural abnormalities in the fetus by ultrasound scanning.

Common recessive genes

Worldwide, the haemoglobinopathies are the most common autosomal recessive disorders. In certain populations, 1 in 6 people are carriers. In white populations 1 in 10 people carry the C282Y haemochromatosis mutation. One in 400 people are therefore homozygous for this mutation, although only one third to one half have clinical signs owing to iron overload. In northern Europeans the commonest autosomal recessive disorder of childhood is cystic fibrosis. Approximately 1 in 25 of the population are carriers. In one couple out of every 625, both partners will be carriers, resulting in an incidence of about 1 in 2500 for cystic fibrosis.

Variability

Autosomal recessive disorders usually demonstrate full penetrance and little clinical variability within families. Haemochromatosis is unusual in that not all homozygotes develop clinical disease. Women in particular are protected by menstruation. In childhood onset SMA type I (Werdnig–Hoffman disease) there is very little difference in the age at death between affected siblings. However, the age at onset, severity and age at death is more variable in intermediate SMA type II. Variation in the severity of an autosomal recessive disorder between families is generally explained by the specific mutation present in the gene. In cystic fibrosis, delta F508 is the most common mutation and most affected homozygotes have pancreatic insufficiency. Patients with other particular mutations are more likely to be pancreatic sufficient, may have less severe pulmonary disease if the regulatory function of the gene is preserved, or even present with just congenital absence of the vas deferens.

New mutations

New mutations are rare in autosomal recessive disorders and it can generally be assumed that both parents of an affected child are carriers. New mutations have occasionally been documented and occur in about 1% of SMA type I cases, where a child inherits a mutation from one carrier parent with a new mutation arising in the gene inherited from the other, non-carrier parent. Recurrence risks for future siblings is therefore very low.

Uniparental disomy

Occasionally, autosomal recessive disorders can arise through a mechanism called uniparental disomy, in which a child inherits two copies of a particular chromosome from one parent and none from the other. If the chromosome inherited in this uniparental fashion carries an autosomal recessive gene mutation, then the child will be an affected homozygote. Recurrence risk for future siblings is extremely low.

Heterogeneity

Genetic heterogeneity is common and involves multiple alleles at a single locus as well as multiple loci for some disorders. Allelic heterogeneity implies that many different mutations can occur in a disease gene. It is common for affected individuals to

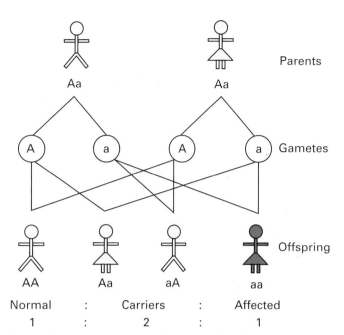

Figure 6.6 Segregation of autosomal recessive alleles when both parents are carriers

Box 6.4 Examples of autosomal recessive disorders

Congenital adrenal hyperplasia
Cystic fibrosis
Deafness (some forms)
Friedreich ataxia
Galactosaemia
Haemochromatosis
Homocystinuria
Hurler syndrome (MPS I)
Oculocutaneous albinism
Phenylketonuria
Sickle cell disease
Tay–Sachs disease
Thalassaemia

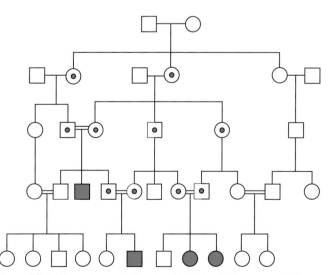

Figure 6.7 Pedigree demonstrating the effect of multiple consanguinity on the inheritance of an autosomal recessive disorder. Affected children (■; ●) have been born to several couples and the obligate gene carriers are indicated (□; ○)

have two different mutations in the disease-causing gene and these people are referred to as compound heterozygotes. The severity of the disorder may be influenced by the particular combination of mutations present. Locus heterogeneity, where a particular phenotype can be caused by different genes, is seen in some autosomal recessive disorders. A number of recessive genes at different loci cause severe congenital deafness and this affects recurrence risk when two affected individuals have children (see chapter 8).

Consanguinity

Consanguinity increases the risk of a recessive disorder because both parents are more likely to carry the same defective gene, that has been inherited from a common ancestor. The rarer the condition the more likely it is to occur when the parents were related before marriage. Overall, the increased risk of having a child with severe abnormalities, including recessive disorders, is about 3% above the risk in the general population.

X linked recessive inheritance

In X linked recessive conditions males are affected because they have only a single copy of genes carried by the X chromosome (hemizygosity), but the disorder can be transmitted through healthy female carriers. A female carrier of an X linked recessive disorder will transmit the condition to half her sons, and half her daughters will be carriers. An unaffected male does not transmit the disorder. An affected male will transmit the mutant gene to all his daughters (who must inherit his X chromosome), but to none of his sons (who must inherit his Y chromosome). This absence of male to male transmission is a hallmark of X linked inheritance. Many X linked recessive disorders are severe or lethal during early life, however, so that the affected males do not reproduce.

Affected females

Occasionally a heterozygous female will show some features of the condition and is referred to as a manifesting carrier. This is usually due to non-random X inactivation leading to the chromosome that carries the mutant allele remaining active in most cells. The process of X inactivation that occurs in early embryogenesis is normally random, so that most female carriers would have around 50% of the normal gene remaining active, which is sufficient to prevent clinical signs. Alternatively, X chromosome abnormalities such as Turner syndrome may give rise to X linked disorders in females since, like males, they are hemizygous for genes carried by the X chromosome. The homozygous affected state may occur in females whose father is affected and whose mother is a carrier. This is only likely to occur in common X linked disorders such as red-green colour blindness, or glucose-6-phosphate dehydrogenase deficiency in the Middle East.

Detecting carriers

Recognising X linked recessive inheritance is important because many women in the family may be at risk of being carriers and of having affected sons, irrespective of whom they marry. Genetic assessment is important because of the high recurrence risk and the severity of many X linked disorders. An X linked recessive condition must be considered when the family history indicates maternally related affected males in different generations of the family. Family history is not always positive, however, as new mutations are common, particularly in conditions that are lethal in affected males. Identifying female gene carriers requires interpretation of the family pedigree and the results of specific carrier tests, including direct mutation detection, DNA linkage studies or biochemical analysis. These

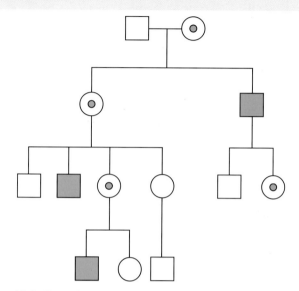

Figure 6.8 Pedigree demonstrating X linked recessive inheritance

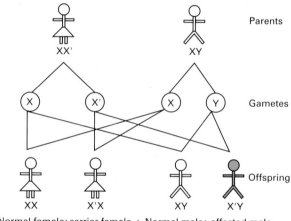

Normal female: carrier female : Normal male: affected male
1 : 1 : 1 : 1

Figure 6.9 Segregation of an X linked recessive allele when the mother is a carrier

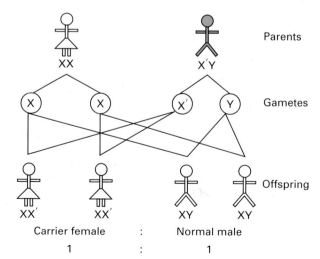

Carrier female : Normal male
1 : 1

Figure 6.10 Segregation of an X linked recessive allele when the father is affected

are discussed further in chapter 9. Carrier detection is not always straightforward as the mothers of some isolated cases may have normal carrier test results but carry germline mutations leading to a risk of recurrence. In cases where mutation analysis cannot be undertaken, biochemical tests and/or linkage analyses are often possible, but may not give definitive results.

> **Box 6.7 Characteristics of X linked recessive inheritance**
> - Males affected almost exclusively
> - Transmission through carrier females
> - Male to male transmission does not occur
> - All daughters of affected males are carriers

> **Box 6.6 Examples of X linked recessive disorders**
> - Anhidrotic ectodermal dysplasia
> - Becker muscular dystrophy
> - Choroideraemia
> - Colour blindness
> - Duchenne muscular dystrophy
> - Emery–Dreifuss muscular dystrophy
> - Fabry disease
> - Fragile X syndrome
> - G-6-P-D deficiency
> - Haemophilia A, B
> - Hunter syndrome (MPS II)
> - Ichthyosis (steroid sulphatase deficiency)
> - Lesch–Nyhan syndrome
> - Menkes syndrome
> - Ocular albinism
> - Ornitine transcarbamylase deficiency
> - Retinitis pigmentosa (some types)
> - Testicular feminisation syndrome

X linked dominant inheritance

An X linked dominant gene will give rise to a disorder that affects both hemizygous males and heterozygous females. Although dominant, females may be less severely affected than males, as in X linked hypophosphataemia (vitamin D-resistant rickets) and oculomotor nystagmus, because of X inactivation which results in expression of the mutant allele in only a proportion of cells. The gene is transmitted through families in the same way as X linked recessive genes: females transmit the mutation to half their sons and half their daughters; males transmit the mutation to all their daughters and none of their sons. The pedigree, however, resembles autosomal dominant inheritance except that there is no male to male transmission and there is an excess of affected females. In some disorders the condition appears to be lethal in affected males, for example focal dermal hypoplasia (Goltz syndrome) and incontinentia pigmenti. In these families there will be fewer males than expected, half of the females will be affected and all surviving males will be unaffected. An affected woman therefore has a one in three chance of having an affected child and two thirds of her children will be girls. Rett syndrome is a disorder that affects girls almost exclusively and usually occurs sporadically, since affected females do not reproduce. This disorder has been shown to be due to a mutation in a gene located at Xq24, confirming that it is an X linked dominant condition.

Y linked inheritance

In Y linked inheritance, only males would be affected, with transmission being from a father to all his sons via the Y chromosome. This pattern of inheritance has previously been suggested for such conditions as porcupine skin, hairy ears, and webbed toes. In most conditions in which Y linked inheritance has been postulated the actual mode of inheritance is probably autosomal dominant with sex limitation. Genes involved in male development and spermatogenesis are carried by the Y chromosome, but the mode of inheritance is not demonstrated because of the associated infertility.

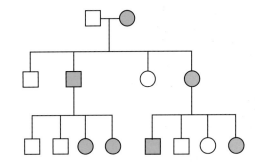

Figure 6.11 Pedigree demonstrating X linked dominant inheritance

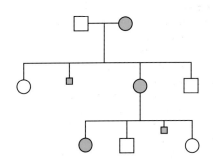

Figure 6.12 Pedigree demonstrating X linked dominant inheritance with lethality in males

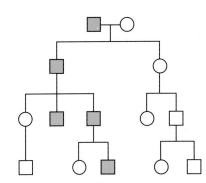

Figure 6.13 Pedigree demonstrating Y linked inheritance

7 Unusual inheritance mechanisms

Unstable mutations

It has generally been assumed that mutant alleles causing mendelian disorders are transmitted unchanged from parent to child. In 1991 the discovery of unstable trinucleotide repeat expansion mutations identified a novel genetic mechanism underlying a number of important disorders.

Several genes are known to contain regions of trinucleotide repeats. The number of repeats varies from person to person in the general population, but within the normal range these repeats are stably transmitted. When the number of repeats is increased beyond the normal range, this region becomes unstable with a tendency to increase in size when transmitted to offspring. In some conditions there is a clear distinction between normal and pathological alleles. In others, the expanded alleles may act either as premutations or as full pathological mutations. Premutations do not cause disease but are unstable and likely to expand further when transmitted to offspring. Once the repeat reaches a certain size it becomes a full mutation and disease will occur. Since the age at onset and severity of the disease correlate with the size of the expansion, this phenomenon accounts for the clinical anticipation that is seen in this group of conditions, where age at onset decreases in successive generations of a family. There is a sex bias in the transmission of the most severe forms of some of these disorders, with maternal transmission of congenital myotonic dystrophy and fragile X syndrome, but paternal transmission of juvenile Huntington disease.

A number of late onset neurodegenerative disorders (for example Huntington disease and spinocerebellar ataxias) are associated with expansions of a CAG repeat sequence in the coding region of the relevant gene, that is translated into polyglutamine tracts in the protein product. These mutations confer a specific gain of function and cause the protein to form intranuclear aggregates that result in cell death. There is usually a clear distinction between normal- and disease-causing alleles in the size of their respective number of repeats and no other types of mutation are found to cause these disorders.

In other disorders (for example fragile X syndrome and Friedreich ataxia) very large expansions occur, which prevent transcription of the gene, and act recessively as loss of function mutations. Other types of mutations occur occasionally in these genes resulting in the same phenotype. In myotonic dystrophy the pathological mechanism of the expanded repeat is not known. It is likely that the expansion affects the transcriptional process of several neighbouring genes. Juvenile myoclonus epilepsy is due to the expansion of a longer repeat region (CCCCGCCCCGCG) normally present in two to three copies in the gene promoter region, expanding to 40 or more repeats in mutant alleles. Trinucleotide repeat expansions have also been found in other conditions (for example polysyndactyly from HOXD13 mutation), where the pathological expansion shows no instability.

Uniparental disomy

Another unusual mechanism causing human disease is that of uniparental disomy (UPD), where both copies of a particular chromosome are inherited from one parent and none from the other. Usually, UPD arises by loss of a chromosome from a conception that was initially trisomic (trisomy rescue). The resulting zygote could contain one chromosome from each

Table 7.1 Trinucleotide repeat expansions

Gain of function mutations (due to CAG repeat)		Normal repeat number	Pathological repeat number
Huntington disease (HD)		6–35	36–100+
Kennedy syndrome (SBMA)		9–35	38–62
Spinocerebellar ataxias	SCA 1	6–38	39–83
	SCA 2	14–31	32–77
(Machado–Joseph	SCA 3	12–39	62–86
disease)	SCA 6	4–17	21–30
	SCA 7	7–35	37–200
Dentatorubro-pallidoluysian atrophy		3–35	49–88

Table 7.2 Trinucleotide repeat expansions

Loss of function mutations	Repeat sequence	Normal repeat number	Unstable repeat number
Fragile XA (site A)	CGG	6–50	55–1000+
Fragile XE (site E)	CCG	6–52	55+
Friedreich ataxia (FA)	GAA	7–22	200–1700
Myotonic dystrophy	CTG	33–35	50–4000+
Spinocerebellar ataxia 8	CTG	16–37	110–500+

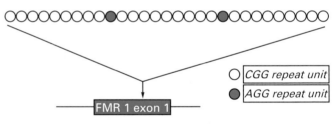

Figure 7.1 5′ end of the *FMR1* gene showing the critical repeat region expanded in fragile X syndrome

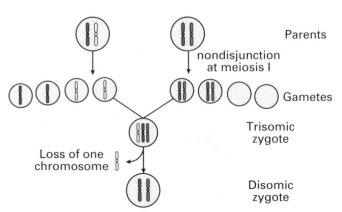

Figure 7.2 Uniparental disomy (heterodisomy) due to nondisjunction at meiosis I

parent (normal), two identical chromosomes from one parent (isodisomy) or two different chromosomes from one parent (heterodisomy). Occasionally UPD may arise by fertilisation of a monosomic gamete followed by duplication of the chromosome from the other gamete (monosomy rescue). This mechanism results in uniparental isodisomy. Theoretically, UPD could also arise by fertilisation of a momosomic gamete with a disomic gamete, resulting in either isodisomy or heterodisomy.

Uniparental disomy may have no clinical consequence by itself. It is occasionally detected by the unmasking of a recessive disorder for which only one parent is a carrier when there is isodisomy for the parental chromosome carrying such a mutation. In this rare situation the child would be affected by a recessive disorder for which the other parent is not a carrier. Recurrence risk for the disorder in siblings is extremely low since UPD is not likely to occur again in another pregnancy.

The other situation in which UPD will have an effect is when the chromosome involved contains one or more imprinted genes, as described in the next section.

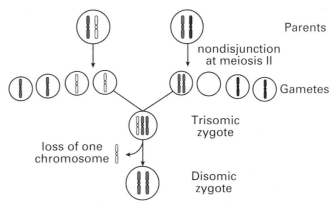

Figure 7.3 Uniparental disomy (isodisomy) due to nondisjunction at meiosis II

Imprinting

It has been observed that some inherited traits do not conform to the pattern expected of classical mendelian inheritance in which genes inherited from either parent have an equal effect. The term imprinting is used to describe the phenomenon by which certain genes function differently, depending on whether they are maternally or paternally derived. The mechanism of DNA modification involved in imprinting remains to be explained, but it confers a functional change in particular alleles at the time of gametogenesis determined by the sex of the parent. The imprint lasts for one generation and is then removed, so that an appropriate imprint can be re-established in the germ cells of the next generation.

The effects of imprinting can be observed at several levels: that of the whole genome, that of particular chromosomes or chromosomal segments, and that of individual genes. For example, the effect of triploidy in human conceptions depends on the origin of the additional haploid chromosome set. When paternally derived, the placenta is large and cystic with molar changes and the fetus has a large head and small body. When the extra chromosome set is maternal, the placenta is small and underdeveloped without cystic changes and the fetus is noticeably underdeveloped. An analogous situation is seen in conceptions with only a maternal or paternal genetic contribution. Androgenic conceptions, arising by replacement of the female pronucleus with a second male pronucleus, give rise to hydatidiform moles which lack embryonic tissues. Gynogenetic conceptions, arising by replacement of the male pronucleus with a second female one, results in dermoid cysts that develop into multitissue ovarian teratomas.

One of the best examples of imprinting in human disease is shown by deletions in the q11-13 region of chromosome 15, which may cause either Prader–Willi syndrome or Angelman syndrome. The features of Prader–Willi syndrome are severe neonatal hypotonia and failure to thrive with later onset of obesity, behaviour problems, mental retardation, characteristic facial appearance, small hands and feet and hypogonadism. Angelman syndrome is quite distinct and is associated with severe mental retardation, microcephaly, ataxia, epilipsy and absent speech.

Prader–Willi and Angelman syndromes are caused by distinct genes within the 15q11-13 region that are subject to different imprinting. The Prader–Willi gene is only active on the chromosome inherited from the father and the Angelman

Figure 7.4 Blonde hair and characteristic facial appearance of Prader–Willi syndrome in child with good weight control, normal intellectual development and minimal behavioral problems

Figure 7.5 Ataxic gait in child with Angelman syndrome

syndrome gene is only active on the chromosome inherited from the mother. Similar de novo cytogenetic or molecular deletions can be detected in both conditions. Prader–Willi syndrome occurs when the deletion affects the paternally derived chromosome 15, whereas the Angelman syndrome occurs when it affects the maternally derived chromosome. In most patients with Prader–Willi syndrome who do not have a chromosome deletion, both chromosome 15s are maternally derived (uniparental disomy). When UPD involves imprinted regions of the genome it has the same effect as a chromosomal deletion arising from the opposite parental chromosome. In Prader–Willi syndrome both isodisomy (inheritance of identical chromosome 15s from one parent) and heterodisomy (inheritance of different 15s from the same parent) have been observed. Uniparental disomy is rare in Angelman syndrome, but when it occurs it involves disomy of the paternal chromosome 15. Other cases are due to mutations within the Angelman syndrome gene (*UBE3A*) that affect its function.

Imprinting has been implicated in other human diseases, for example familial glomus tumours that occur only in people who inherit the mutant gene from their father and Beckwith–Wiedemann syndrome that occurs when maternally transmitted.

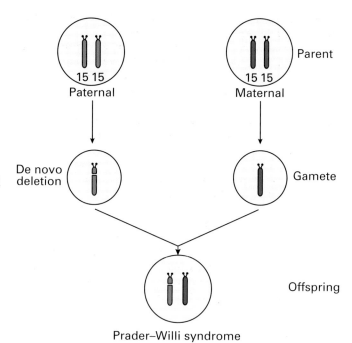

Figure 7.6 Prader–Willi syndrome in offspring as a consequence of a de novo deletion affecting the paternally transmitted chromosome 15

Mosaicism

Mosaicism refers to the presence of two or more cell lines in an individual that differ in chromosomal constitution or genotype, but have been derived from a single zygote. Mosaicism may involve whole chromosomes or single gene mutations and is a postzygotic event that arises in a single cell. Once generated, the genetic change is transmitted to all daughter cells at cell division, creating a second cell line. The process can occur during early embryonic development, or in later fetal or postnatal life. The time at which the mosaicism develops will determine the relative proportions of the two cell lines, and hence the severity of the phenotype caused by the abnormal cell line. Chimaeras have a different origin, being derived from the fusion of two different zygotes to form a single embryo. Chimaerism explains the rare occurrence of both XX and XY cell lines in a single individual.

Functional mosaicism occurs in all females as only one X chromosome remains active in each cell. The process of X inactivation occurs in early embryogenesis and is random. Thus, alleles that differ between the two chromosomes will be expressed in mosaic fashion. Carriers of X linked recessive mutations normally remain asymptomatic as only a proportion of cells have the mutant allele on the active chromosome. Occasional females will, by chance, have the normal X chromosome inactivated in the majority of cells and will then manifest systemic symptoms of the disorder caused by the mutant gene. In X linked dominant disorders such as incontinentia pigmenti, female gene carriers have patchy skin pigmentation that follows Blaschko's lines because of the mixture of normal and mutant cells in the skin during development.

Chromosomal mosaicism is not infrequent, and arises by postzygotic errors in mitosis. Mosaicism is observed in conditions such as Turner syndrome and Down syndrome, and the phenotype is less severe than in cases with complete aneuploidy. Mosaicism has been documented for many other numerical or structural chromosomal abnormalities that would be lethal in non-mosaic form. The clinical importance of chromosomal mosaicism detected prenatally may be difficult to assess. The abnormal karyotype detected by amniocentesis or chorionic villus sampling may be confined to placental cells,

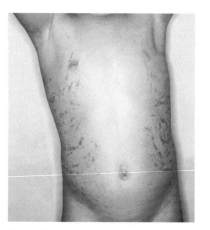

Figure 7.7 Patchy distribution of skin lesions in female with incontinentia pigmenti, an X linked dominant disorder, lethal in males but not in females, because of functional X chromosomal mosaicism (courtesy of Professor Dian Donnai, Regional Genetic Service, St Mary's Hospital Manchester)

Figure 7.8 Tetrasomy for chromosome 12p occurs only in mosaic form in liveborn infants (extra chromosome composed of two copies of the short arm of chromosome 12 arrowed) (courtesy of Dr Lorraine Gaunt and Helena Elliott, Regional Genetic service, St Mary's Hospital, Manchester)

but even when present in the fetus the severity with which the fetus will be affected is difficult to predict.

Single gene mutations occurring in somatic cells also result in mosaicism. In mendelian disorders this may present as a patchy phenotype, as in segmental neurofibromatosis type 1. Somatic mutation is also a mechanism responsible for neoplastic change.

Germline mosaicism is one explanation for the transmission of a genetic disorder to more than one offspring by apparently normal parents. In these cases the mutation may be confined to the germline cells or may be present in a proportion of somatic cells as well. In Duchenne muscular dystrophy, it has been calculated that up to 20% of the mothers of isolated cases, whose carrier tests performed on leucocyte DNA give normal results, may have gonadal mosaicism for the muscular dystrophy mutation. The possibility of germline mosaicism makes it difficult to exclude a risk of recurrence in other X linked recessive disorders where the mother's carrier tests give normal results, and autosomal dominant disorders where the parents are clinically unaffected.

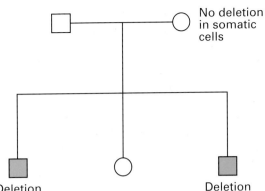

No deletion in somatic cells

Deletion   Deletion

Figure 7.9 Pedigree showing recurrence of Duchenne muscular dystrophy due to dystrophin gene deletion in the sons of a woman who does not carry the deletion in her leucocyte DNA. Recurrence is caused by gonadal mosaicism, in which the mutation is confined to some of the germline cells in the mother

Mitochondrial disorders

Not all DNA is contained within the cell nucleus. Mitochondria have their own DNA consisting of a double-stranded circular molecule. This mitochondrial DNA consists of 16 569 base pairs that constitute 37 genes. There is some difference in the genetic code between the nuclear and mitochondrial genomes, and mitochondrial DNA is almost exclusively coding, with the genes containing no intervening sequences (introns). A diploid cell contains two copies of the nuclear genome, but there may be thousands of copies of the mitochondrial genome, as each mitochondrion contains up to 10 copies of its circular DNA and each cell contains hundreds of mitochondria. The mitochondrial genome encodes 22 types of transfer and two ribosomal RNA molecules that are involved in mitochondrial protein synthesis, as well as 13 of the polypeptides involved in the respiratory chain complex. The remaining respiratory chain polypeptides are encoded by nuclear genes. Diseases affecting mitochondrial function may therefore be controlled by nuclear gene mutation and follow mendelian inheritance, or may result from mutations within the mitochondrial DNA.

Mutations within mitochondrial DNA appear to be 5 or 10 times more common than mutations in nuclear DNA, and the

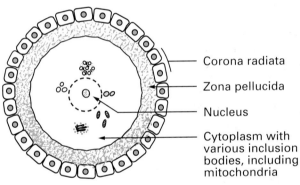

Corona radiata

Zona pellucida

Nucleus

Cytoplasm with various inclusion bodies, including mitochondria

Figure 7.10 Representation of human egg

Table 7.3 Examples of diseases caused by mitochondrial DNA mutations

Disorder	Symptoms	Common mutation	Inheritance
Leber hereditary optic neuropathy (LHON)	Acute visual loss and possibly other neurological symptoms	Point mutation at position 11778 in ND4 gene of complex 1	Maternal
MERRF	Myoclonic epilepsy, other neurological symptoms and ragged red fibres in skeletal muscle	Point mutation in tRNA-Lys gene (position 8344)	Maternal
Kaerns–Sayre syndrome	Progressive external ophthalmoplegia, pigmentary retinopathy, heart block, ataxia, muscle weakness, deafness	Large deletion (position 8470-13447) Large tandem duplication	Usually sporadic Sporadic
MELAS	Encephalomyopathy, lactic acidosis, stroke-like episodes	Point mutation in tRNA-Leu gene (position 3243)	Maternal

accumulation of mitochondrial mutations with time has been suggested as playing a role in ageing. As the main function of mitochondria is the synthesis of ATP by oxidative phosphorylation, disorders of mitochondrial function are most likely to affect tissues such as the brain, skeletal muscle, cardiac muscle and eye, which contain abundant mitochondria and rely on aerobic oxidation and ATP production. Mutations in mitochondrial DNA have been identified in a number of diseases, notably Leber hereditary optic neuropathy (LHON), myoclonic epilepsy with ragged red fibres (MERRF), mitochondrial myopathy with encephalopathy, lactic acidosis, and stroke-like episodes (MELAS), and progressive external ophthalmoplegia including Kaerns–Sayre syndrome.

Disorders due to mitochondrial mutations often appear to be sporadic. When they are inherited, however, they demonstrate maternal transmission. This is because only the egg contributes cytoplasm and mitochondria to the zygote. All offspring of a carrier mother may carry the mutation, all offspring of a carrier father will be normal. The pedigree pattern in mitochondrial inheritance may be difficult to recognise, however, because some carrier individuals remain asymptomatic. In Leber hereditary optic neuropathy, which causes sudden and irreversible blindness, for example, half the sons of a carrier mother are affected, but only 1 in 5 of the daughters become symptomatic. Nevertheless, all daughters transmit the mutation to their offspring. The descendants of affected fathers are unaffected.

Because multiple copies of mitochondrial DNA are present in the cell, mitochondrial mutations are often heteroplasmic – that is, a single cell will contain a mixture of mutant and wild-type mitochondrial DNA. With successive cell divisions some cells will remain heteroplasmic but others may drift towards homoplasmy for the mutant or wild-type DNA. Large deletions, which make the remaining mitochondrial DNA appreciably shorter, may have a selective advantage in terms of replication efficiency, so that the mutant genome accumulates preferentially. The severity of disease caused by mitochondrial mutations probably depends on the relative proportions of wild-type and mutant DNA present, but is very difficult to predict in a given subject.

Box 7.1 Genetic counselling dilemmas in mitochondrial diseases

- Some disorders of mitochondrial function are due to nuclear gene mutations
- Some disorders caused by mitochondrial mutations are sporadic
- When maternally transmitted, not all offspring are affected
- Severity is very variable and difficult to predict
- Prenatal diagnosis is not feasible

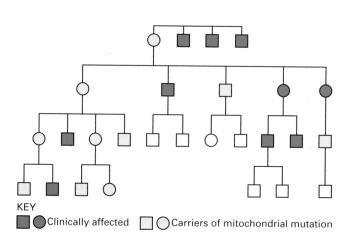

KEY

■ ●Clinically affected   □ ○Carriers of mitochondrial mutation

Figure 7.11 Pedigree of Leber hereditary optic neuropathy caused by a mutation within the mitochondrial DNA. Carrier women transmit the mutation to all their offspring, some of whom will develop the disorder. Affected or carrier men do not transmit the mutation to any of their offspring

8 Estimation of risk in mendelian disorders

This chapter gives some examples of simple risk calculations in mendelian disorders. Risks may be related to the probability of a person developing a disorder or to the probability of transmitting it to their offspring. Mathematical risk calculated from the pedigree data may often be modified by additional information, such as biochemical test results. In an increasing number of disorders, gene carriers can be identified with certainty by gene mutation analysis. Risk calculation remains important, since decisions about whether to proceed with a genetic test are often influenced by the level of risk determined from the pedigree. Risks or probabilities are usually expressed in terms of a percentage (i.e. 25%) a fraction (i.e. 1/4 or 1 in 4) or as odds (i.e. 3 to 1 against or 1 to 3 in favour) of a particular outcome.

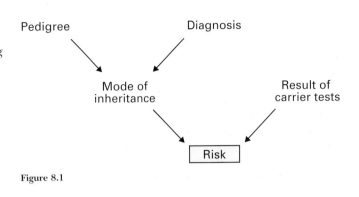

Figure 8.1

Autosomal dominant disorders

Examples 1–4
Many autosomal dominant disorders have onset in adult life and are not apparent clinically during childhood. In such families a clinically unaffected adolescent or young adult has a high risk of carrying the gene, but an unaffected elderly relative is unlikely to do so. The prior risk of 50% for developing the disorder can therefore be modified by age. Data are available for Huntington disease (Harper PS and Newcombe RG, *J Med Genet* 1992; **29**: 239–42) from which age-related risks can be derived for clinically unaffected relatives. In example 1 the risk of developing Huntington disease for individual B is still almost 50% at the age of 30. Risk to offspring C is therefore 25%. In example 2, individual B remains unaffected at the age of 60 and her residual risk is reduced to around 20%. Risk to offspring C at the age of 40 is reduced to around 5% after his own age-related risk adjustment. In example 3 the risk to B is reduced to 6% at the age of 70 and the risk to the 40-year old son is less than 2%. In example 4 the risk for C at the age of 40 is only reduced to around 17%, because parent B, although clinically unaffected, died aged 30 while still at almost 50% risk.

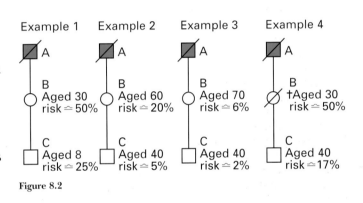

Figure 8.2

Example 5
When both parents are affected by the same autosomal dominant disorder the risk of having affected children is high, as shown in example 5. The chance of a child being unaffected is only 1 in 4. The risk of a child being an affected heterozygote is 1 in 2 and of being an affected homozygote is 1 in 4. In most conditions, the phenotype in homozygous individuals is more severe than that in heterozygotes, as seen in familial hypercholesterolaemia and achondroplasia. In some disorders, such as Huntington disease and myotonic dystrophy, the homozygous state is not more severe and this probably reflects the mode of action of the underlying gene mutation.

When both parents are affected by different autosomal dominant disorders, the chance of a child being unaffected by either condition is again 1 in 4. The risk of being affected by one or other condition is 1 in 2 and the risk of inheriting both conditions is 1 in 4.

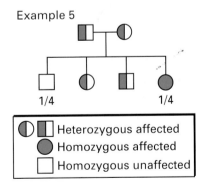

Figure 8.3

Example 6
Reduced penetrance also modifies simple autosomal dominant risk. Reduced penetrance refers to the situation in which not all carriers of a particular dominant gene mutation will develop

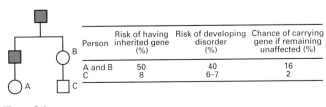

Figure 8.4

35

clinical signs or symptoms. Genes demonstrating reduced penetrance include tuberous sclerosis, retinoblastoma and otosclerosis. Example 6 shows the risk to the child and grandchild of an affected individual for a disorder with 80% penetrance in which only 80% of gene mutation carriers develop the disorder. Although clinically unaffected, individuals A and B may still carry the mutant gene. The risk to individual C is small. In general the risk of clinical disease affecting the grandchild of an affected person is fairly low if the intervening parent is unaffected. The maximum risk does not exceed 10% since disorders with low penetrance are unlikely to cause disease and disorders with high penetrance are unlikely to be transmitted by an unaffected parent.

Many autosomal dominant disorders show variable expression, with different degrees of disease severity being observed in different people from the same family. Although the risk of offspring being affected is 50%, the family may be more concerned to know the likelihood of severe disease occurring. The incidence of severe manifestations or disease complications has been documented for many autosomal disorders, such as neurofibromatosis type 1, and these figures can be used in counselling. For example, around 10% of people with Charcot–Marie–Tooth disease type 1 (CMT1) have severe difficulties with ambulation by the age of 40 years. An affected individual therefore has a 5% risk overall for having a child who will become severely disabled.

Autosomal recessive disorders

Example 7
Recurrence of autosomal recessive disorders generally occurs only within one particular sibship in a family. Occurrence of the same disorder in different sibships within an extended family can occur if the mutant gene is common in the population, or there is multiple consanguinity. Many members of the family will, however, be gene carriers and may wish to know the risk for their own children being affected. Example 7 shows the risk for relatives being carriers in a family where an autosomal recessive disorder has occurred, ignoring the possibility that both partners in a particular couple may be carriers apart from the parents of the affected child.

Example 8
The risk of an unaffected sibling having an affected child is low and is determined by the chance that their partner is also a carrier. The actual risk depends on the frequency of the mutant gene in the population. This can be calculated from the disease incidence using the Hardy–Weinberg equilibrium principle. In general, doubling the square root of the disease incidence gives a sufficiently accurate estimation of carrier frequency in a given population. The risk for cystic fibrosis is shown in example 8. The unaffected sibling of a person with cystic fibrosis has a carrier risk of 2/3. The unrelated spouse has the population risk of around one in 22 for being a carrier. Since the risk of both parents passing on the mutant gene is one in four if they are both carriers, the risk to their child would be $2/3 \times 1/22 \times 1/4$.

Example 9
When there is a tradition of consanguinity, more than one marriage between related individuals may occur in a family. If a consanguineous couple have a child affected by an autosomal recessive condition other marriages within the family may be at increased risk for the same condition. The risk can be defined by calculating the carrier risk for both partners as shown in example 9. Marriage within the family may be an important cultural factor

Table 8.1		
Disease	**Complication**	**Risk (%)**
Neurofibromatosis 1	Learning disability:	
	mild	30
	moderate–severe	3
	Malignancy	5
	Scoliosis	10
Tuberous sclerosis	Epilepsy	60
	Learning disability: (moderate–severe)	40
Myotonic dystrophy	Severe congenital onset when maternally transmitted	20
Waardenburg syndrome 1	Deafness	25

Example 7

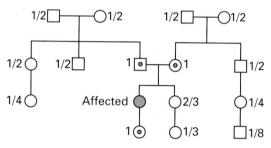

Figure 8.5

Example 8

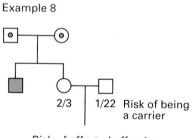

2/3 | 1/22 Risk of being a carrier

Risk of affected offspring
$2/3 \times 1/22 \times 1/4 = 1/132$

Figure 8.6

Example 9

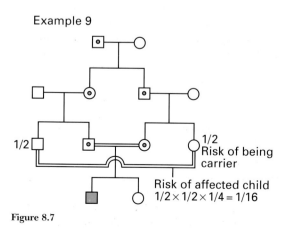

1/2 | 1/2 Risk of being carrier

Risk of affected child
$1/2 \times 1/2 \times 1/4 = 1/16$

Figure 8.7

and the risk of an autosomal recessive disorder may not influence choice of a marital partner. If carrier tests are possible for a condition that has occurred in the family, testing may provide reassurance, or identify couples whose pregnancies will be at risk, and for whom prenatal diagnosis might be appropriate.

Example 10

When an affected person has children, the risk of recurrence is again determined by the chance that the partner is a carrier. In non-consanguineous marriages this is calculated from the population carrier frequency. In consanguineous marriages it is calculated from degree of the relationship to the spouse. The affected parent must pass on a gene for the disorder since they are homozygous for this gene and the risk to the offspring is therefore half of the spouse's carrier risk (the chance that they too would pass on a mutant gene). The risk in a consanguineous family is shown in example 10.

Examples 11 and 12

Some autosomal recessive disorders, such as severe congenital deafness can be caused by a variety of genes at different loci. When both parents are affected by autosomal recesive deafness, the risks to the offspring will depend on whether the parents are homozygous for the same (allelic) or different (non-allelic) genes. In example 11 both parents have the same form of recessive deafness and all their children will be affected. In example 12 the parents have different forms of recessive deafness due to genes at separate loci. Their offspring will be heterozygous at both loci, but not affected by deafness. Since the different types of autosomal deafness cannot always be identified by genetic testing at present, the risk to offspring in this situation cannot be clarified until the presence or absence of deafness in the first-born child is known.

Example 13

Twin pregnancies complicate the estimation of recurrence risk. In monozygous twins, both will be either affected or unaffected. The risk that both will be affected is 25%, as with singleton pregnancies. In dizygous twins, however, it is possible that only one twin or that both twins might be affected. Example 13 shows the risks for one, or both, being affected by an autosomal recessive disorder when the zygosity is known (dizygous) or unknown. When zygosity is unknown the risks are calculated using the relative frequencies of monozygosity (1/3) and dizygosity (2/3).

X linked recessive disorders

Example 14

Calculation of risks in X linked recessive disorders is important since many female relatives may have a substantial carrier risk although they are usually completely healthy, and carriers have a high risk of transmitting the disorder irrespective of whom they marry. Calculation of risks is often complex and requires referral to a specialist genetic centre. Risks are determined by combining information from pedigree structure and the results of specific tests. If there is more than one affected male in a family, certain female relatives who are obligate carriers can be identified. Example 14 shows a pedigree identifying a number of obligate and potential carriers, indicating the risks to several other female relatives.

Examples 15 and 16

Since a carrier has a 50% chance of transmitting the condition to each of her sons, it follows that a woman who has several unaffected but no affected sons is less likely to be a carrier. This information can be used to modify a woman's prior risk of

Example 10

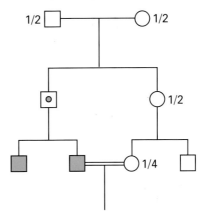

Risk of affected child
$1 \times 1/4 \times 1/2 = 1/8$

Figure 8.8

Example 11 | Example 12

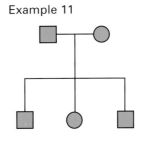
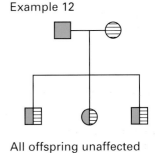

All offspring affected | All offspring unaffected

Figure 8.9

Example 13

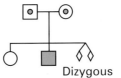
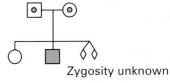

Dizygous | Zygosity unknown

	Dizygous	Zygosity unknown
Only one affected	37.5%	25%
Both affected	6.25%	12.5%
Neither affected	56.25%	62.5%

Figure 8.10

Exampe 14

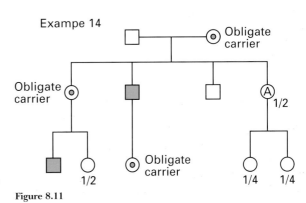

Figure 8.11

being a carrier using Bayesian calculation methods. Details of this are given in a number of specialised texts listed in the bibliography, including Young ID. Introduction to risk calculation genetic counselling. Oxford University Press 1991. Examples 15 and 16 indicate how the carrier risk for individual A from example 14 can be reduced if she has one unaffected son or four unaffected sons, without going into details of the actual calculation.

Example 17

In lethal X linked recessive disorders new mutations account for a third of all cases. When there is only one affected boy in a family, his mother is therefore not always a carrier. Carrier risks in families with an isolated case of such a disorder (for example Duchenne muscular dystrophy) are shown in example 17. These risks can be modified by molecular analysis if the underlying mutation in the affected boy can be identified, or by serum creatine kinase levels in the female relatives. Gonadal mosaicism is common in the mothers of isolated cases of Duchenne muscular dystrophy, occurring in around 20% of mothers whose somatic cells show no gene mutation, so that recurrence risk is not negligible.

Isolated cases

Example 18

Pedigrees showing only one affected person are the type most commonly encountered in clinical practice, since many cases present after the first affected family member is diagnosed (as in example 18). Various causes must be considered, and risk estimation in this situation depends entirely on reaching an accurate diagnosis in the affected person. In conditions amenable to molecular genetic diagnosis, such as Charcot–Marie–Tooth disease and Becker muscular dystrophy, mutation detection enables provision of definite risks to family members. In other cases, probabilities calculated from pedigree data cannot be made more certain.

There are several explanations to account for isolated cases of an autosomal dominant disorder. These include new mutation and non-paternity. Recurrence risks are negligible unless one parent is a non-penetrant gene carrier or has a mutation restricted to germline cells. Autosomal and X linked recessive disorders usually present after the birth of the first affected child. Recurrence risks are high unless an X linked disorder is due to a new mutation. The recurrence risks for most chromosomal disorders are low, the exception being those due to a balanced chromosome rearrangement in one parent (see chapters 4 and 5). Disorders with a polygenic or multifactorial aetiology often have relatively low recurrence risks. Studies documenting recurrence in the families of affected individuals provide data on which to base empiric recurrence risks. Some of these disorders are discussed in Chapter 12.

Example 19

In some disorders there are both genetic and non-genetic causes. If these cannot be distinguished by clinical features or specific investigations, calculation of risk needs to be based on the relative frequency of the different causes. In isolated cases of severe congenital deafness, for example, it is estimated that 70% of cases are genetic, once known environmental causes have been excluded. Of the genetic cases, around two thirds follow autosomal recessive inheritance. The calculation of recurrence risk after an isolated case of severe congenital deafness is shown in example 20.

Example 15   Example 16

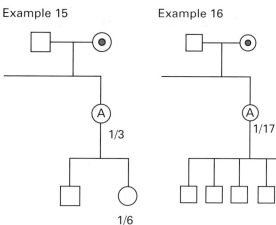

Figure 8.12   Figure 8.13

Example 17

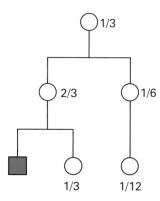

Figure 8.14

Example 18

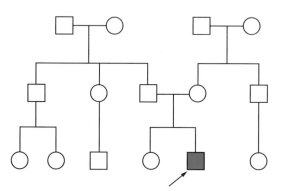

Figure 8.15

> **Box 8.1 Possible causes of sporadic cases**
> - Autosomal dominant
> - Autosomal recessive
> - X linked recessive
> - Chromosomal
> - Polygenic (multifactorial)
> - Non-genetic

Example 19

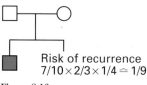

Risk of recurrence
$7/10 \times 2/3 \times 1/4 \simeq 1/9$

Figure 8.16

9   Detection of carriers

Identifying carriers of genetic disorders in families or populations at risk plays an important part in preventing genetic disease. A carrier is a healthy person who possesses the mutant gene for an inherited disorder in the heterozygous state, which they may transmit to their offspring. The implications for themselves and their offspring depend on whether the gene mutation acts in a dominant or recessive fashion. In recessive disorders gene carriers remain unaffected, but in late onset dominant conditions, gene carriers will be destined to develop the condition themselves at some stage. Autosomal recessive gene mutations are extremely common and everyone carries at least one gene for a recessive disorder and one or more that would be lethal in the homozygous state. However, an autosomal recessive gene transmitted to offspring will be of consequence only if the other parent is also a carrier and transmits a mutant gene as well. Whenever dominant or X linked recessive gene mutations are transmitted, however, the offspring will be affected.

The term carrier is generally restricted to people at risk of transmitting mendelian disorders and does not apply to parents whose children have chromosomal abnormalities such as Down syndrome or congenital malformations such as neural tube defects. An exception is that people who have balanced chromosomal translocations are referred to as carriers, as the inheritance of balanced or unbalanced translocations follows mendelian principles.

Obligate carriers

In families in which there is a genetic disorder some members must be carriers because of the way in which the condition is inherited. These obligate carriers can be identified by drawing a family pedigree and they do not require testing as their genetic state is not in doubt. Obligate carriers of autosomal dominant, autosomal recessive and X linked disorders are shown in the box. Identifying obligate carriers is important not only for their own counselling but also for defining a group of individuals in whom tests for carrier state can be evaluated. When direct mutation analysis is not possible, information is needed regarding the proportion of obligate carriers who show abnormalities on clinical examination or with specific investigations, to enable interpretation of carrier test results in possible carriers. In late onset autosomal dominant disorders it is also important to know at what age obligate carriers develop signs of the condition so that appropriate advice can be given to relatives at risk.

Autosomal dominant disorders

In autosomal dominant conditions most heterozygous subjects are clinically affected and testing for carrier state applies only to disorders that are either variable in their manifestations or have a late onset. Gene carriers in conditions such as tuberous sclerosis may be minimally affected but run the risk of having severely affected children, whereas carriers in other disorders, such as Huntington disease, are destined to develop severe disease themselves.

Identifying asymptomatic gene carriers allows a couple to make informed reproductive decisions, may indicate a need to avoid environmental triggers (as in porphyria or malignant hyperthermia), or may permit early treatment and prevention

Box 9.1 Risks to offspring of carriers
- Recessive mutations: no risk unless partner is a carrier
- Dominant mutations: 50% risk applies to late onset disorders
- X linked recessive: 50% risk to male offspring

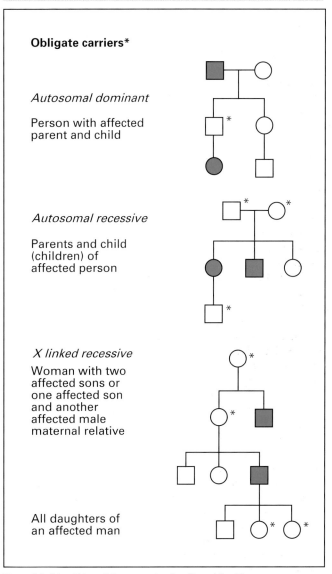

Figure 9.1 Identifying obligate carriers in affected families

Box 9.2 Some autosomal dominant disorders amenable to carrier detection
- Adult polycystic kidney disease
- Charcot–Marie–Tooth disease
- Facioscapulohumeral dystrophy
- Familial adenomatous polyposis
- Familial breast cancer (*BRCA 1* and *2*)
- Familial hypercholesterolaemia
- Huntington disease
- Malignant hyperthermia
- Myotonic dystrophy
- Porphyria
- Spinocerebellar ataxia
- Tuberous sclerosis
- von Hippel–Lindau disease

of complications (as in von Hippel–Lindau disease and familial adenomatous polyposis). Although testing for carrier state has important benefits in conditions in which the prognosis is improved by early detection, it is also possible in conditions not currently amenable to treatment such as Huntington disease and other late onset neurodegenerative disorders. It is crucial that appropriate counselling and support is available before predictive tests for these conditions are undertaken, as described in chapter 3. Exclusion of carrier state is a very important aspect of testing, since this relieves anxiety about transmitting the condition to offspring and removes the need for long term follow up.

Autosomal recessive disorders

In autosomal recessive conditions carriers remain healthy and carrier testing is done to define risks to offspring. Occasionally, heterozygous subjects may show minor abnormalities, such as altered red cell morphology in sickle cell disease and mild anaemia in thalassaemia.

The parents of an affected child can be considered to be obligate carriers. New mutations and uniparental disomy are very rare exceptions where a child is affected when only one parent is a carrier. The parents of an affected child do not need testing unless this is to determine the underlying mutation to allow prenatal diagnosis when there are no surviving affected children.

For the healthy siblings and other relatives of an affected person, carrier testing for themselves and their partners is only appropriate if the condition is fairly common or they are consanguineous. Testing for carrier state in the relatives of an individual with an autosomal recessive disorder is referred to as cascade screening. This type of testing is offered by some centres for cystic fibrosis. The clinical diagnosis of cystic fibrosis in a child is confirmed by mutation analysis of the *CFTR* gene. If the child has two different mutations, the parents are tested to see which mutation they each carry. Relatives can then be tested for the appropriate mutation to see if they are carriers or not. For those shown to be carriers, their partners can then be tested. Since there are over 700 mutations that have been described in the *CFTR* gene, partners are tested only for the most common mutations in the appropriate population. If no mutation is detected, their carrier risk can be reduced from their 1 in 25 population risk to a very low level, although not absolutely excluded. In this situation, the risk of cystic fibrosis affecting future offspring is very small and prenatal diagnosis is not indicated. The main reason for offering cascade screening is to identify couples where both partners are carriers before they have an affected child. In these cases, prenatal diagnosis is both feasible and appropriate.

In rare recessive conditions there is little need to test relatives since their partners are very unlikely to be carriers for the same condition. In many cases it is possible to do carrier tests on a family member by testing for the mutation present in the affected relative. However, it is seldom helpful to identify the family member as a carrier if the partner's carrier state cannot be determined. It is more important to calculate and explain the risk to their offspring, which is usually sufficiently low to be reassuring and to remove the need for prenatal diagnosis.

X linked recessive disorders

Carrier detection in X linked recessive conditions is particularly important as these disorders are often severe, and in an affected

---

Box 9.3 Some autosomal recessive disorders amenable to carrier detection

Population-based screening
- Thalassaemia
- Tay–Sachs disease
- Sickle cell disease
- Cystic fibrosis

Family-based testing*
- Alpha 1-antitrypsin deficiency
- Batten disease
- Congenital adrenal hyperplasia
- Friedreich ataxia
- Galactosaemia
- Haemochromatosis
- Mucopolysaccharidosis 1 (Hurler syndrome)
- Phenylketonuria
- Spinal muscular atrophy (SMA I, II, and III)

*Indicated or feasible in families with an affected member

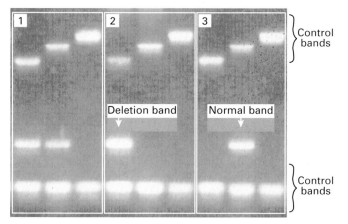

Figure 9.2 Analysis of ΔF508 mutation status in cystic fibrosis using ARMS analysis.
Panel 1: ΔF508 heterozygote – the sample shows both deletion-specific and normal bands
Panel 2: ΔF508 homozygote – the sample shows only the deletion-specific band and no normal band
Panel 3: Normal control – the sample shows only a normal band indicating the absence of the ΔF508 mutation

---

Box 9.4 Some X linked recessive disorders amenable to carrier detection

- Adrenoleucodystrophy
- Albinism (ocular)
- Alport syndrome
- Angiokeratoma (Fabry disease)
- Choroideraemia
- Chronic granulomatous disease
- Ectodermal dysplasia (anhidrotic)
- Fragile X syndrome
- Glucose-6-phosphate dehydrogenase deficiency
- Haemophilia A and B
- Ichthyosis (steroid sulphatase deficiency)
- Lesch–Nyhan syndrome
- Menkes syndrome
- Mucopolysaccharidosis II (Hunter syndrome)
- Muscular dystrophy (Duchenne and Becker)
- Ornithine transcarbarmylase deficiency
- Retinitis pigmentosa
- Severe combined immune deficiency (SCID)

family many female relatives may be carriers at risk of having affected sons, irrespective of whom they marry. Genetic counselling cannot be undertaken without accurate assessment of carrier state, and calculating risks is often complex.

In families with more than one affected male, obligate carriers can be identified and prior risks to other female relatives calculated. A variety of tests can then be used to determine carrier state and to undertake prenatal diagnosis. In families with only one affected male, the situation regarding genetic risk is more complex, because of the possibility of new mutation. New mutations are particularly frequent in severe conditions such as Duchenne muscular dystrophy and may arise in several ways. One third of cases arise by new mutation in the affected boy, with only two thirds of mothers of isolated cases being carriers. If the boy has inherited the mutation from his mother, she may carry the mutation in mosaic form, limited to the germline cells, in which case other female relatives will not be at increased risk. Alternatively, the mutation may represent a new event occurring when the mother was conceived, or a mutation transmitted to her by her mother or occasionally her father, which might be present in other female relatives.

Obligate carriers of X linked disorders do not always show abnormalities on biochemical testing because of lyonisation, a process by which one or other X chromosome in female embryos is randomly inactivated early in embryogenesis. The proportion of cells with the normal or mutant X chromosome remaining active varies and will influence results of carrier tests. Carriers with a high proportion of normal X chromosomes remaining active will show no abnormalities on biochemical testing. Conversely, carriers with a high proportion of mutant X chromosomes remaining active are more likely to show biochemical abnormalities and may occasionally develop signs and symptoms of the disorder. Females with symptoms are called manifesting carriers.

Biochemical tests designed to determine carrier state must be evaluated initially in obligate carriers identified from affected families. Only tests which give significantly different results in obligate carriers compared with controls will be useful in determining the genetic state of female relatives at risk. Because the ranges of values in obligate carriers and controls overlap considerably (for example serum creatine kinase activity in X linked muscular dystrophy) the results for possible carriers are expressed in relative terms as a likelihood ratio. With this type of test, confirmation of carrier state is always easier than exclusion. In muscular dystrophy a high serum creatine kinase activity confirms the carrier state but a normal result does not eliminate the chance that the woman is a carrier.

The problem of lyonisation can be largely overcome if biochemical tests can be performed on clonally derived cells. Hair bulbs have been successfully used to detect carriers of Hunter syndrome (mucopolysaccharidosis II). Carriers can be identified because they have two populations of hair bulbs, one with normal iduronate sulphatase activity, reflecting hair bulbs with the normal X chromosome remaining active, and the other with low enzyme activity, representing those with the mutant X chromosome remaining active.

DNA analysis is not affected by lyonisation and is the method of choice for detecting carriers. Initial analysis using linked or intragenic probes is being replaced by more direct testing as mutation analysis becomes feasible. When direct mutation testing is not possible, calculating the probability of carrier state entails analysis of pedigree data with the results of linkage analysis and other tests. The calculation employs Bayesian analysis, and computer programs are available for the complex analysis required in large families. The possibility of new mutation and gonadal mosaicism in the mother must be

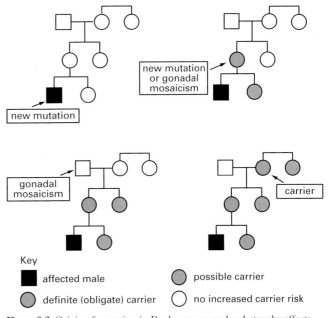

Key

■ affected male   ● possible carrier

● definite (obligate) carrier   ○ no increased carrier risk

Figure 9.3 Origin of mutation in Duchenne muscular dystrophy affects carrier risks within families

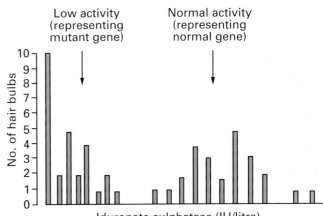

Figure 9.4 Two populations of hair bulbs with low and normal activity of iduronate sulphatase, respectively, in female carrier of Hunter syndrome

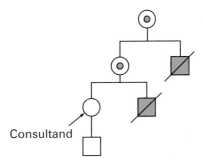

Information on consultand (relative at risk):

Prior risk = 50% (mother obligate carrier)

Risk modified by:

DNA linkage analysis – reducing prior risk to 5%
One healthy son – reducing risk
Analysis of serum creatine kinase activity – giving probability of carrier state of 0.3

Risk after Bayesian calculation = 1%

Figure 9.5 Calculation of carrier risk in Duchenne muscular dystrophy where the underlying mutation is not known

taken into account in sporadic cases. In the case of gonadal mosaicism the results of carrier tests will be normal in the mother of the affected boy.

Methods of testing

Various methods of testing can be used to determine carrier state, including physical examination, physiological and biochemical tests, imaging and molecular genetic analysis. Tests related directly to gene structure and function discriminate better than those measuring secondary biochemical consequences of the mutant gene. Detection of a secondary abnormality may confirm the carrier state but its apparent absence does not always guarantee normality.

Clinical signs

Careful examination for clinical signs may identify some carriers and is particularly important in autosomal dominant conditions in which the underlying biochemical basis of the disorder is unknown or where molecular analysis is not routinely available, as in Marfan syndrome and neurofibromatosis type 1. In some X linked recessive disorders, especially those affecting the eye or skin, abnormalities may be detected by clinical examination in female carriers. The absence of clinical signs does not exclude carrier state.

Clinical examination can be supplemented with investigations such as physiological studies, microscopy and radiology, for example: nerve conduction studies in Charcot–Marie–Tooth disease, electroretinogram in retinitis pigmentosa and renal scan in adult polycystic kidney disease. In myotonic dystrophy, before direct mutation analysis became possible, asymptomatic carriers could usually be identified in early adult life by a combination of clinical examination to detect myotonia and mild weakness of facial, sternomastoid and distal muscles, slit lamp examination of the eyes to detect lens opacities, and electromyography to look for myotonic discharges. Presymptomatic genetic testing can now be achieved by molecular analysis, but clinical examination is still important, since early clinical signs may be apparent, indicating that a genetic test is likely to give a positive result. Some people may decide not to go ahead with a definitive genetic test in this situation. Confirmation or exclusion of the carrier state is important for genetic counselling, especially for mildly affected women who have an appreciable risk of producing severely affected infants with the congenital form of myotonic dystrophy.

Analysis of genes

DNA analysis has revolutionised testing for genetic disorders and can be applied to both carrier testing in autosomal and X linked recessive disorders and to predictive testing in late onset autosomal dominant disorders. The genes for most important mendelian disorders are now mapped and many have been cloned. Direct mutation analysis is possible for an increasing number of conditions. This provides definitive results for carrier tests, presymptomatic diagnosis, and prenatal diagnosis when a pathological mutation is detected. If a mutation cannot be detected, linkage analysis within affected families may still be possible, contributing to carrier detection and prenatal diagnosis. Methods of DNA analysis and its application to genetic disease are discussed in later chapters.

In some disorders all carriers have the same mutation. Examples include the point mutation in sickle cell disease and the trinucleotide repeat expansions in Huntington disease and myotonic dystrophy. In these cases, carrier detection by

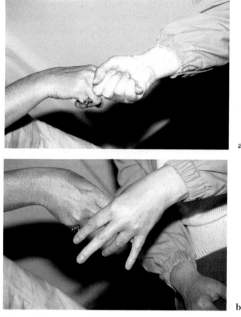

Figure 9.6 a) and b) Myotonia of grip is one of the first signs detected in myotonic dystrophy

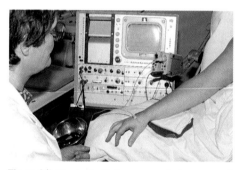

Figure 9.7 Myotonic discharges on electromyography may be demonstrated in the absence of clinical signs in myotonic dystrophy

---

Box 9.5 Examples of some common mendelian disorders amenable to carrier or presymptomatic testing by direct mutation analysis

Carrier testing
- haemoglobinopathies
- cystic fibrosis
- Duchenne muscular dystrophy
- Fragile X syndrome
- Spinal muscular atrophy (SMA I, II, III)

Presymptomatic testing
- Huntington disease
- Myotonic dystrophy
- Spinocerebellar ataxia (types 1,2,3,6,7,8,12)
- Charcot–Marie–Tooth disease (type 1)
- Familial adenomatous polyposis

molecular analysis is straightforward. In most genetic disorders, however, there are a large number of different mutations that can occur in the gene responsible for the condition. In these disorders, identifying the mutation (or mutations) present in the affected individual enables carrier status of relatives to be determined with certainty but it is not usually possible to determine carrier status in an unrelated spouse. At best, exclusion of the most common mutations in the spouse will reduce their carrier risk in comparison to the general population risk.

For conditions where mutation analysis is not possible, or does not identify the mutation underlying the disorder, carrier testing in relatives may still be possible using linked DNA markers to track the disease gene through the family. This approach will not identify carrier status in unrelated spouses, so is mainly applicable to autosomal dominant or X linked conditions and only appropriate for autosomal recessive disorders if there is consanguinity.

Analysis of gene products

When DNA analysis is not feasible, biochemical identification of carriers may be possible when the gene product is known. This approach can be used for some inborn errors of metabolism caused by enzyme deficiency as well as for disorders caused by a defective structural protein, such as haemophilia and thalassaemia. Overlap between the ranges of values in heterozygous and normal people occurs even when the primary gene product is being analysed, and interpretation of results can be difficult.

Secondary biochemical abnormalities

When the gene product is not known or cannot be readily tested, the identification of carriers may depend on detecting secondary biochemical abnormalities. Raised serum creatine kinase activity in some carriers of Duchenne and Becker muscular dystrophies has been a very useful carrier test and is still used in conjunction with linkage analysis when the underlying mutation cannot be identified. The overlap between the ranges of values in normal subjects and gene carriers is often considerable, and the sensitivity of this type of test is only moderate. Abnormal test results make carrier state highly likely, but normal results do not necessarily indicate normality.

Population screening

The main opportunity for preventing recessive disorders depends on population screening programmes, which identify couples at risk before the birth of an affected child within the family. Screening tests must be sufficiently sensitive to avoid false negative results and yet specific enough to avoid false positive results. To be employed on a large scale the tests must also be safe, simple and fairly inexpensive. In addition, screening programmes need to confer benefits to individual subjects as well as to society, and stigmatisation must be avoided if they are to be successful.

Population screening aimed at identifying carriers of common autosomal recessive disorders allows the identification of carrier couples before they have an affected child, and provides the opportunity for first trimester prenatal diagnosis. Carrier screening programmes for thalassaemia and Tay–Sachs disease in high risk ethnic groupings in several countries have resulted in a significant reduction in the birth prevalence of

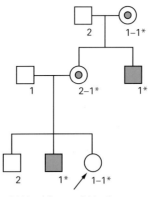

1, 2 = DNA variants within the Duchenne muscular dystrophy gene on X chromosome

Figure 9.8 Prediction of carrier state by detecting intragenic DNA variations in Duchenne muscular dystrophy. Disease gene cosegregates with DNA variant 1*, predicting that consultand (↗) is a carrier

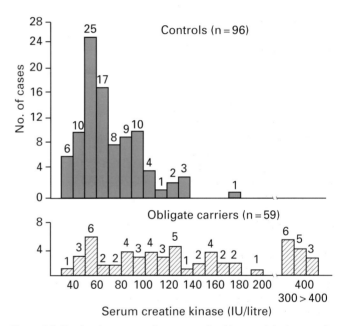

Figure 9.9 Overlapping ranges of serum creatine kinase activity in controls and obligate carriers of Becker muscular dystrophy. (Ranges vary among laboratories.)

Figure 9.10 Collecting a mouth wash sample for DNA extraction and carrier testing for cystic fibrosis

these disorders. Carrier screening for sickle cell disease has been less successful. Carrier screening for cystic fibrosis is also possible, although not all carriers can be identified because of the diversity of mutations within the cystic fibrosis gene. Screening programmes instituted in antenatal clinics and in general practice have reported a substantial uptake for cystic fibrosis carrier testing when it is offered, but indicate that few couples actively seek this type of test themselves. It is important that appropriate information and counselling is available to individuals being offered screening, as they are likely to have little prior knowledge of the disorder being screened for and the implications of a positive test result. Specific training will be needed by members of primary health care and obstetric teams before any new screening programmes are instituted, as these are the settings in which such tests are likely to be offered.

In addition to screening programmes aimed at identifying carriers, there are well established programmes for screening all neonates to identify those affected by conditions such as phenylketonuria and hypothyroidism, where early diagnosis and treatment is successful in preventing mental retardation. The value of including other metabolic disorders in screening programmes depends on the incidence of the disorder and the prospect of altering the prognosis by its early detection. Possible candidates include galactosaemia, maple syrup urine disease and congenital adrenal hyperplasia.

> **Box 9.6 Conditions amenable to population screening programmes**
>
> **Antenatal**
> - Thalassaemia
> - Sickle cell disease
> - Tay–Sachs disease
> - Cystic fibrosis
>
> **Neonatal**
> - Phenylketonuria
> - Hypothyroidism
> - Galactosaemia

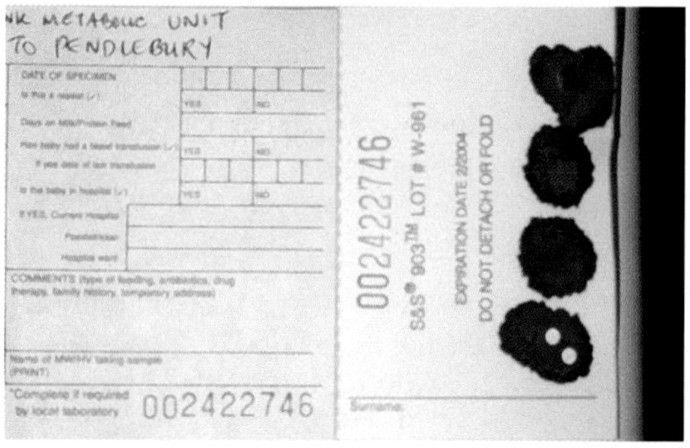

Figure 9.11 Neonatal blood samples used for biochemical screening

10 Single gene disorders

There are thousands of genetic traits and disorders described, some of which are exceedingly rare. All of the identified mendelian traits in man have been catalogued by McKusick and are listed on the Omim (online mendelian inheritance in man) database described in chapter 16. In this chapter the clinical and genetic aspects of a few examples of some of the more common disorders are briefly outlined and examples of genetic disorders affecting various organ systems are listed. Molecular analysis of some of these conditions is described in chapter 18.

Central nervous system disorders

Huntington disease

Huntington disease is an autosomal dominant disease characterised by progressive choreiform movements, rigidity, and dementia from selective, localised neuronal cell death associated with atrophy of the caudate nucleus demonstrated by CNS imaging. The frequency of clinical disease is about 6 per 100 000 with a frequency of heterozygotes of about 1 per 10 000. Development of frank chorea may be preceded by a prodromal period in which there are mild psychiatric and behavioural symptoms. The age of onset is often between 30 and 40 years, but can vary from the first to the seventh decade. The disorder is progressive, with death occurring about 15 years after onset of symptoms. Surprisingly, affected homozygotes are not more severely affected than heterozygotes and new mutations are exceedingly rare. Clinical treatment trials commenced in 2000 to assess the effect of transplanting human fetal striatal tissue into the brain of patients affected by Huntington disease as a potential treatment for neurodegenerative disease.

The gene (designated *IT15*) for Huntington disease was mapped to the short arm of chromosome 4 in 1983, but not finally cloned until 1993. The mutation underlying Huntington disease is an expansion of a CAG trinucleotide repeat sequence (see chapter 7). Normal alleles contain 9–35 copies of the repeat, whereas pathological alleles usually contain 37–86 repeats, but sometimes more. Transcription and translation of pathological alleles results in the incorporation of an expanded polyglutamine tract in the protein product (huntingtin) leading to accumulation of intranuclear aggregates and neuronal cell death. Clinical severity of the disorder correlates with the number of trinucleotide repeats. Alleles that contain an intermediate number of repeats do not always cause disease and may not be fully penetrant. Instability of the repeat region is more marked on paternal transmission and most cases of juvenile onset Huntington disease are inherited from an affected father.

Prior to the identification of the mutation, presymptomatic predictive testing could be achieved by linkage studies if the family structure was suitable. Prenatal testing could also be undertaken. In some cases tests were done in such a way as to identify whether the fetus had inherited an allele from the clinically affected grandparent without revealing the likely genetic status of the intervening parent. This enabled adults at risk to have children predicted to be at very low risk without having predictive tests themselves. Direct mutation detection now enables definitive confirmation of the diagnosis in clinically affected individuals (see chapter 18) as well as providing presymptomatic predictive tests and prenatal diagnosis. Considerable experience has been gained with

Table 10.1 Examples of autosomal dominant adult-onset diseases affecting the central nervous system for which genes have been cloned

Disease		Gene
Familial alzheimer disease	AD1	amyloid precursor gene (*APP*)
	AD2	*APOE*4* association
	AD3	Presenilin-1 gene (*PSEN 1*)
	AD4	Presenilin-2 gene (*PSEN 2*)
Familial amyotrophic lateral sclerosis ALS1		superoxide dismutase-1 gene (*SOD1*)
ALS susceptibility		heavy neurofilament subunit gene (*NEFH*)
Familial Parkinson disease PARK1 +lewy body PARK4		alpha-synuclein gene (*SNCA*)
Frontotemporal dementia with Parkinsonism		microtubule-associated protein tau gene (*MAPT*)
Creutzfeldt-Jakob disease (CJD)		prion protein gene (*PRNP*)
Cerebral autosomal dominant arteriopathy with subcortical infarcts and leucoencephalopathy(CADASIL)		*NOTCH 3*
Familial British dementia (FBD)		*ITM2B*

Box 10.1 Neurological disorders due to trinucleotide repeat expansion mutations

Huntington disease (HD)
Fragile X syndrome (FRAXA)
Fragile X site E (FRAXE)
Kennedy syndrome (SBMA)
Myotonic dystrophy (DM)
Spinocerebellar ataxias (SCA 1,2,6,7,8,12)
Machado-Joseph disease (SCA3)
Dentatorubral-pallidolysian atrophy (DRPLA)
Friedreich ataxia (FA)
Oculopharyngeal muscular dystrophy (OPMD)

Table 10.2 Inheritance pattern and gene product for some common neurological disorders

Disorder	Inheritance	Gene product
Childhood onset spinal muscular atrophy	AR	SMN protein
Kennedy syndrome (SBMA)	XLR	androgen receptor
Myotonia congenita (Thomsen type)	AD	muscle chloride channel
Myotonia congenita (Becker type)	AD	muscle chloride channel
Friedreich ataxia	AR	frataxin
Spinocerebellar ataxia type 1	AD	ataxin-1
Charcot–Marie–Tooth type 1a	AD	peripheral myelin protein P22
Charcot–Marie–Tooth type 1b	AD	peripheral myelin protein zero
Hereditary spastic paraplegia (SPG4)	AD	spastin
Hereditary spastic paraplegia (SPG7)	AR	paraplegin
Hereditary spastic paraplegia (SPG2)	XLR	propeolipid protein

predictive testing and an agreed protocol has been drawn up for use in clinical practice that is applicable to other predictive testing situations (see chapter 3).

Fragile X syndrome

Fragile X syndrome, first described in 1969 and delineated during the 1970s, is the most common single cause of inherited mental retardation. The disorder is estimated to affect around 1 in 4000 males, with many more gene carriers. The clinical phenotype comprises mental retardation of varying degree, macro-orchidism in post-pubertal males, a characteristic facial appearance with prominent forehead, large jaw and large ears, joint laxity and behavioural problems.

Chromosomal analysis performed under special culture conditions demonstrates a fragile site near the end of the long arm of the X chromosome in most affected males and some affected females, from which the disorder derived its name. The disorder follows X linked inheritance, but is unusual because of the high number of female carriers who have mental retardation and because there is transmission of the gene through apparently unaffected males to their daughters – a phenomenon not seen in any other X linked disorders. These observations have been explained by the nature of the underlying mutation, which is an expansion of a CGG trinucleotide repeat in the *FMR1* gene. Normal alleles contain up to 45 copies of the repeat. Fragile X mutations can be divided into premutations (50–199 repeats) that have no adverse effect on phenotype and full mutations (over 200 repeats) that silence gene expression and cause the clinical syndrome. Both types of mutations are unstable and tend to increase in size when transmitted to offspring. Premutations can therefore expand into full mutations when transmitted by an unaffected carrier mother. All of the boys and about half of the girls who inherit full mutations are clinically affected. Mental retardation is usually moderate to severe in males, but mild to moderate in females. Males who inherit the premutation are unaffected and usually transmit the mutation unchanged to their daughters who are also unaffected, but at risk of having affected children themselves.

Molecular analysis confirms the diagnosis of fragile X syndrome in children with learning disability, and enables detection of premutations and full mutations in female carriers, premutations in male carriers and prenatal diagnosis (see chapter 18).

Neuromuscular disorders

Duchenne and Becker muscular dystrophies

Duchenne and Becker muscular dystrophies are due to mutations in the X linked dystrophin gene. Duchenne muscular dystrophy (DMD) is one of the most common and severe neuromuscular disorders of childhood. The incidence of around 1 in 3500 male births has been reduced to around 1 in 5000 with the advent of prenatal diagnosis for high risk pregnancies.

Boys with DMD may be late in starting to walk. If serum creatine kinase estimation is included as part of the investigations at this stage, very high enzyme levels will indicate the need for further investigation. In the majority of cases, onset of symptoms is before the age of four. Affected boys present with an abnormal gait, frequent falls and difficulty climbing steps. Toe walking is common, along with pseudohypertrophy of calf muscles. Pelvic girdle weakness results in the characteristic waddling gait and the Gower manoeuvre (a manoeuvre by which affected boys use their

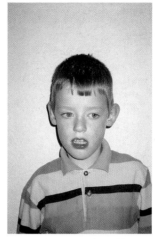

Figure 10.1 Boy with fragile X syndrome showing characteristic facial features: tall forehead, prominent ears and large jaw

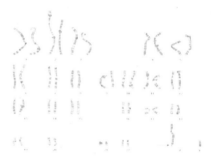

Figure 10.2 Karyotype of a male with fragile X syndrome demonstrating the fragile site on the X chromosome (courtesy of Dr Lorraine Gaunt and Helena Elliott, Regional Genetic Service, St Mary's Hospital, Manchester)

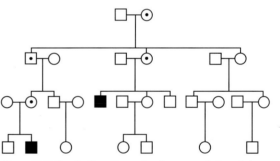

Figure 10.3 Fragile X pedigree showing transmission of the mutation through an unaffected male (⊙ ⊡ premutation carrier, ■ full mutation)

Figure 10.4 Scapular winging, mild lordosis and enlarged calves in the early stages of Duchenne muscular dystrophy

hands to "climb up" their legs to get into a standing position when getting up from the floor). Calf pain is a common symptom at this time. Scapular winging is the first sign of shoulder girdle involvement and, as the disease progresses, proximal weakness of the arm muscles becomes apparent. Most boys are confined to a wheelchair by the age of 12. Flexion contractures and scoliosis are common and require active management. Cardiomyopathy and respiratory problems occur and may necessitate nocturnal respiratory support. Survival beyond the age of 20 is unusual. Intellectual impairment is associated with DMD, with 30% of boys having an IQ below 75.

The diagnosis of DMD is confirmed by muscle biopsy with immunocytochemical staining for the dystrophin protein. Two thirds of affected boys have deletions or duplications within the dystrophin gene that are readily detectable by molecular testing (see chapter 18). The remainder have point mutations that are difficult to detect. Mutation analysis or linkage studies enable carrier detection in female relatives and prenatal diagnosis for pregnancies at risk. However, one third of cases arise by new mutation. Gonadal mosaicism, with the mutation being confined to germline cells, occurs in about 20% of mothers of isolated cases. In these women, the mutation is not detected in somatic cells when carrier tests are performed, but there is a risk of having another affected son. Prenatal diagnosis should therefore be offered to all mothers of isolated cases. Testing for inherited mutations in other female relatives does give definitive results and prenatal tests can be avoided in those relatives shown not to be carriers.

About 5% of female carriers manifest variable signs of muscle involvement, due to non-random X inactivation that results in the abnormal gene remaining active in the majority of cells. There have also been occasional reports of girls being more severely affected as a result of having Turner syndrome (resulting in hemizygosity for a dystrophin gene mutation) or an X:autosome translocation disrupting the gene at Xp21 (causing inactivation of the normal X chromosome and functional hemizygosity).

Becker muscular dystrophy (BMD) is also due to mutations within the dystrophin gene. The clinical presentation is similar to DMD, but the phenotype milder and more variable. The underlying mutations are commonly also deletions. These mutations differ from those in DMD by enabling production of an internally truncated protein that retains some function, in comparison to DMD where no functional protein is produced.

Myotonic dystrophy

Myotonic dystrophy is an autosomal dominant disorder affecting around 1 in 3000 people. The disorder is due to expansion of a trinucleotide repeat sequence in the 3′ region of the dystrophia myotonica protein kinase (*DMPK*) gene. The trinucleotide repeat is unstable, causing a tendency for further expansion as the gene is transmitted from parent to child. The size of the expansion correlates broadly with the severity of phenotype, but cannot be used predictively in individual situations.

Classical myotonic dystrophy is a multisystem disorder that presents with myotonia (slow relaxation of voluntary muscle after contraction), and progressive weakness and wasting of facial, sternomastoid and distal muscles. Other features include early onset cataracts, cardiac conduction defects, smooth muscle involvement, testicular atrophy or obstetric complications, endocrine involvement, frontal balding, hypersomnia and hypoventilation. Mildly affected late onset cases may have little obvious muscle involvement and present with only cataracts. Childhood onset myotonic dystrophy

Figure 10.5 Young boy with Duchenne muscular dystrophy demonstrating the Gower manoeuvre, rising from the floor by getting onto his hands and feet, then pushing up on his knees

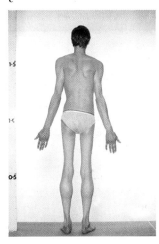

Figure 10.6 Marked wasting of the thighs with calf hypertrophy and scapular winging in young man with Becker muscular dystrophy

Table 10.3 Muscular dystrophies with identified genetic defects

Type of muscular dystrophy	Locus/ gene symbol	Protein deficiency	Inheritance
Congenital	LAMA2	merosin	AR
Congenital	ITGA7	integrin α 7	AR
Duchenne/ Becker	DMD/BMD	dystrophin	XLR
Emery–Dreifuss	EMD	emerin	XLR
Emery–Dreifuss	EDMD-AD	lamin A/C	AD
Facioscapulo-humeral	FSHD	(4q34 rearrangement)	AD
Limb girdle with cardiac involvement	LGMD1B	lamin A/C	AD
Limb girdle	LGMD1C	caveolin-3	AD
	LGMD2A	calpain 3	AR
	LGMD2B	dysferlin	AR
	LGMD2C	γ sarcoglycan	AR
	LGMD2D	α sarcoglycan	AR
	LGMD2E	β sarcoglycan	AR
	LGMD2F	δ sarcoglycan	AR
	LGMD2G	telethonin	AR

usually presents with less specific symptoms of muscle weakness, speech delay and mild learning disability, with more classical clinical features developing later. Congenital onset myotonic dystrophy can occur in the offspring of affected women. These babies are profoundly hypotonic at birth and have major feeding and respiratory problems. Children who survive have marked facial muscle weakness, delayed motor milestones and commonly have intellectual disability and speech delay. The age at onset of symptoms becomes progressively younger as the condition is transmitted through a family. Progression of the disorder from late onset to classical, and then to childhood or congenital onset, is frequently observed over three generations of a family.

Molecular analysis identifies the expanded CTG repeat, confirming the clinical diagnosis and enabling presymptomatic predictive testing in young adults. Prenatal diagnosis is also possible, but does not, on its own, predict how severe the condition is going to be in an affected child.

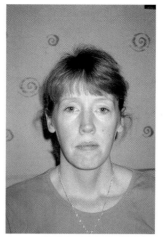

Figure 10.7 Ptosis and facial muscle weakness in a woman with myotonic dystrophy

Neurocutaneous disorders

Neurofibromatosis
Neurofibromatosis type 1 (NF1), initially described by von Recklinghausen, is one of the most common single gene disorders, with an incidence of around 1 in 3000. The main diagnostic features of NF1 are café-au-lait patches, peripheral neurofibromas and lisch nodules. Café-au-lait patches are sometimes present at birth, but often appear in the first few years of life, increasing in size and number. A child at risk who has no café-au-lait patches by the age of five is extremely unlikely to be affected. Freckling in the axillae, groins or base of the neck is common and generally only seen in people with NF1. Peripheral neurofibromas usually start to appear around puberty and tend to increase in number through adult life. The number of neurofibromas varies widely between different subjects from very few to several hundred. Lisch nodules (iris hamartomas) are not visible to the naked eye but can be seen using a slit lamp. Minor features of NF1 include short stature and macrocephaly. Complications of NF1 are listed in the box and occur in about one third of affected individuals. Malignancy (mainly embryonal tumours or neurosarcomas) occur in about 5% of affected individuals. Learning disability occurs in about one third of children, but severe mental retardation in only 1 to 2%.

Clinical management involves physical examination with measurement of blood pressure, visual field testing, visual acuity testing and neurological examination on an annual basis. Children should be seen every six months to monitor growth and development and to identify symptomatic optic glioma and the development of plexiform neurofibromas or scoliosis.

The gene for NF1 was localised to chromosome 17 in 1987 and cloned in 1990. The gene contains 59 exons and encodes of protein called neurofibromin, which appear to be involved in the control of cell growth and differentiation. Mutation analysis is not routine because of the large size of the gene and the difficulty in identifying mutations. Prenatal diagnosis by linkage analysis is possible in families with two or more affected individuals. NF1 has a very variable phenotype and prenatal testing does not predict the likely severity of the condition. Up to one third of cases arise by a new mutation. In this situation,

Box 10.2 Diagnostic criteria for NF1

Two or more of the following criteria:
- Six or more café-au-lait macules
  >5 mm diameter before puberty
  >15 mm diameter after puberty
- Two or more neurofibroma of any type or one plexiform neuroma
- Freckling in the axillary or inguinal regions
- Two or more Lisch nodules
- Optic glioma
- Bony lesions such as pseudarthrosis, thinning of the long bone cortex or sphenoid dysplasia
- First degree relative with NF1 by above criteria

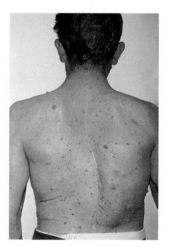

Figure 10.8 Multiple neurofibromas and scoliosis in NF1

Box 10.3 Complications of NF1
- Plexiform neurofibromas
- Congenital bowing of tibia and fibula due to pseudarthrosis
- Optic glioma
- Scoliosis
- Epilepsy
- Hypertension
- Nerve root compression by spinal neurofibromas
- Malignancy
- Learning disability

the recurrence risk is very low for unaffected parents who have had one affected child.

Neurofibromatosis type 2 (NF2) is a disorder distinct from NF1. It is characterised by schwannomas (usually bilateral) and other cranial and spinal tumours. Café-au-lait patches and peripheral neurofibromas can also occur, as in NF1. Survival is reduced in NF2, with the mean age of death being around 32 years. NF2 follows autosomal dominant inheritance with about 50% of cases representing new mutations. The *NF2* gene, whose protein product has been called merlin, is a tumour suppressor gene located on chromosome 22. Mutation analysis of the *NF2* gene contributes to confirmation of diagnosis in clinically affected individuals and enables presymptomatic testing of relatives at risk, identifying those who will require annual clinical and radiological screening.

Tuberous sclerosis complex

Tuberous sclerosis complex (TSC) is an autosomal dominant disorder with a birth incidence of about 1 in 6000. TSC is very variable in its clinical presentation. The classical triad of mental retardation, epilepsy and adenosum sebaceum are present in only 30% of cases. TSC is characterised by hamartomas in multiple organ systems, commonly the skin, CNS, kidneys, heart and eyes. The ectodermal manifestations of the condition are shown in the table. CNS manifestations include cortical tumours that are associated with epilepsy and mental retardation, and subependymal nodules that are found in 95% of subjects on MRI brain scans. Subependymal giant cell astrocytomas develop in about 6% of affected individuals. TSC is associated with both infantile spasms and epilepsy occurring later in childhood. Learning disability is frequently associated. Attention deficit hyperactivity disorder is associated with TSC and severe retardation occurs in about 40% of cases. Renal angiomyolipomas or renal cysts are usually bilateral and multiple, but mainly asymptomatic. Their frequency increases with age. Angiomyolipomas may cause abdominal pain, with or without haematuria, and multiple cysts can lead to renal failure. There may be a small increase in the risk of renal carcinoma in TSC. Cardiac rhabdomyomas are detected by echocardiography in 50% of children with TSC. These can cause outflow tract obstruction or arrhythmias, but tend to resolve with age. Ophthalmic features of TSC include retinal hamartomas, which are usually asymptomatic.

TSC follows autosomal dominant inheritance but has very variable expression both within and between families. Fifty per cent of cases are sporadic. First degree relatives of an affected individual need careful clinical examination to detect minor features of the condition. The value of other investigations in subjects with no clinical features is not of proven benefit.

Two genes causing TSC have been identified: *TSC1* on chromosome 9 and *TSC2* on chromosome 16. The products of these genes have been called hamartin and tuberin respectively. Current strategies for mutation analysis do not identify the underlying mutation in all cases. However, when a mutation is detected, this aids diagnosis in atypical cases, can be used to investigate apparently unaffected parents of an affected child, and enables prenatal diagnosis. Mutations of both *TSC1* and *TSC2* are found in familial and sporadic TSC cases. There is no observable difference in the clinical presentation between *TSC1* and *TSC2* cases, although it has been suggested that intellectual disability is more frequent in sporadic cases with *TSC2* than *TSC1* mutations.

Box 10.4 Diagnostic criteria for NF2
- Bilateral vestibular schwannomas
- First degree relative with NF2 and either
  a) unilateral vestibular schwannoma or
  b) any two features listed below
- Unilateral vestibular schwannoma and two or more other features listed below
- Multiple meningiomas with one other feature listed below

meningioma, glioma, schwannoma, posterior subcapsular lenticular opacities, cerebral calcification

Table 10.4 Some ectodermal manifestations of tuberous sclerosis

Feature	Frequency (%)
Hypomelanotic macule	80–90
Facial angiofibroma (adenosum sebaceum)	80–90
Shagreen patch	20–40
Forehead plaque	20–30
Ungual fibroma 5–14 years	20
>30 years	80
Dental enamel pits	50

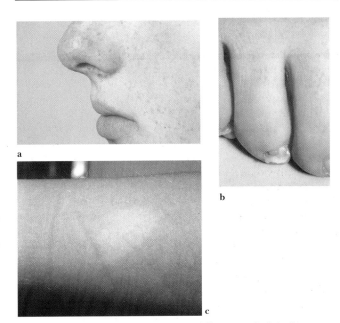

Figure 10.9 Facial angiofibroma, periungal fibroma and ash leaf depigmentation in Tuberous sclerosis

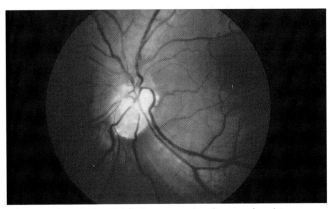

Figure 10.10 Retinal astrocytic hamartoma in tuberous sclerosis (courtesy of Dr Graeme Black, Regional Genetic Service, St Mary's Hospital, Manchester)

Connective tissue disorders

Marfan syndrome

Marfan syndrome is an autosomal dominant disorder affecting connective tissues caused by mutation in the gene encoding fibrillin 1 (*FBN1*). The disorder has an incidence of at least 1 in 10 000. It arises by new mutation in 25–30% of cases. In some familial cases, the diagnosis may have gone unrecognised in previously affected relatives because of mild presentation and the absence of complications.

The main features of Marfan syndrome involve the skeletal, ocular and cardiovascular systems. The various skeletal features of Marfan syndrome are shown in the box. Up to 80% of affected individuals have dislocated lenses (usually bilateral) and there is also a high incidence of myopia. Cardiovascular manifestations include mitral valve disease and progressive dilatation of the aortic root and ascending aorta. Aorta dissection is the commonest cause of premature death in Marfan syndrome. Regular monitoring of aortic root dimension by echocardiography, medical therapy (betablockers) and elective aortic replacement surgery have contributed to the fall in early mortality from the condition over the past 30 years.

Clinical diagnosis is based on the Gent criteria, which require the presence of major diagnostic criteria in two systems, with involvement of a third system. Major criteria include any combination of four of the skeletal features, ectopia lentis, dilatation of the ascending aorta involving at least the sinus of Valsalva, lumbospinal dural ectasia detected by MRI scan, and a first degree relative with confirmed Marfan syndrome. Minor features indicating involvement of other symptoms include striae, recurrent or incisional herniae, and spontaneous pneumothorax.

Clinical features of Marfan syndrome evolve with age and children at risk should be monitored until growth is completed. More frequent assessment may be needed during the pubertal growth spurt. Neonatal Marfan syndrome represents a particularly severe form of the condition presenting in the newborn period. Early death from cardiac insufficiency is common. Most cases are due to new mutations, which are clustered in the same region of the *FBN1* gene. Adults with Marfan syndrome need to be monitored annually with echocardiography. Pregnancy in women with Marfan syndrome should be regarded as high risk and carefully monitored by obstetricians and cardiologists with expertise in management of the condition.

Marfan syndrome was initially mapped to chromosome 15q by linkage studies and subsequently shown to be associated with mutations in the fibrillin 1 gene (*FBN1*). Fibrillin is the major constituent of extracellular microfibrils and is widely distributed in both elastic and non-elastic connective tissue throughout the body. *FBN1* mutations have been found in patients who do not fulfil the full diagnostic criteria for Marfan syndrome, including cases with isolated ectopia lentis, familial aortic aneurysm and patients with only skeletal manifestations. *FBN1* is a large gene containing 65 exons. Most Marfan syndrome families carry unique mutations and more than 140 different mutations have been reported. Screening new cases for mutations is not routinely available, and diagnosis depends on clinical assessment. Mutations in the fibrillin 2 gene (*FBN2*) cause the phenotypically related disorder of contractural arachnodactyly (Beal syndrome) characterised by dolichostenomelia (long slim limbs) with arachnodactyly, joint contractures and a characteristically crumpled ear.

Box 10.5 Skeletal features of Marfan syndrome

Major features
- Thumb sign (thumb nail protrudes beyond ulnar border of hand when adducted across palm)
- Wrist sign (thumb and 5th finger overlap when encircling wrist)
- Reduced upper : lower segment ratio (<0.85)
- Increased span : height ratio (>1.05)
- Pectus carinatum
- Pectus excavatum requiring surgery
- Scoliosis >20° or spondylolisthesis
- Reduced elbow extension
- Pes planus with medical displacement of medial maleolus
- Protrusio acetabulae

Minor features
- Moderate pectus excavatum
- Joint hypermobility
- High arched palate with dental crowding
- Characteristic facial appearance

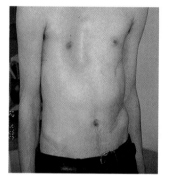

Figure 10.11 Marked pectus excavatum in Marfan syndrome

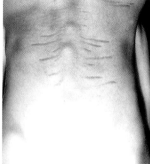

Figure 10.12 Multiple striae in Marfan syndrome

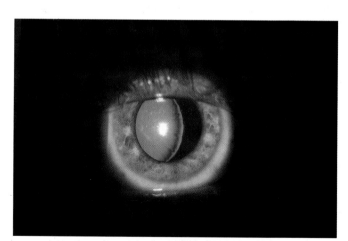

Figure 10.13 Dislocated lenses in Marfan syndrome (courtesy of Dr Graeme Black, Regional Genetic Service, St Mary's Hospital, Manchester)

Cardiac and respiratory disorders

Cystic fibrosis

Cystic fibrosis (CF) is the most common lethal autosomal recessive disorder of childhood in Northern Europeans. The incidence of cystic fibrosis is approximately 1 in 2000, with 1 in 22 people in the population being carriers. Clinical manifestations are due to disruption of exocrine pancreatic function (malabsorption), intestinal glands (meconium ileus), bile ducts (biliary cirrhosis), bronchial glands (chronic bronchopulmonary infection with emphysema), sweat glands (abnormal sweat electrolytes), and gonadal function (infertility). Clinical presentation is very variable and can include any combination of the above features. Some cases present in the neonatal period with meconium ileus, others may not be diagnosed until middle age. Presentation in childhood is usually with failure to thrive, malabsorption and recurrent pneumonia. Approximately 15% of patients do not have pancreatic insufficiency. Congenital bilateral absence of the vas deferens is the usual cause of infertility in males with CF and can occur in heterozygotes, associated with a particular mutation in intron 8 of the gene.

Cystic fibrosis is due to mutations in the cystic fibrosis conductance regulator (*CFTR*) gene which is a chloride ion channel disease affecting conductance pathways for salt and water in epithelial cells. Decreased fluid and salt secretion is responsible for the blockage of exocrine outflow from the pancreas, accumulation of mucus in the airways and defective reabsorption of salt in the sweat glands. Family studies localised the gene causing cystic fibrosis to chromosome 7q31 in 1985 and the use of linked markers in affected families enabled carrier detection and prenatal diagnosis. Prior to this, carrier detection tests were not available and prenatal diagnosis, only possible for couples who already had an affected child, relied on measurement of microvillar enzymes in amniotic fluid – a test that was associated with both false positive and false negative results. Direct mutation analysis now forms the basis of both carrier detection and prenatal tests (see chapter 18).

Newborn screening programmes to detect babies affected by CF have been based on detecting abnormally high levels of immune reactive trypsin in the serum. Diagnosis is confirmed by a positive sweat test and *CFTR* mutation analysis. Within affected families, mutation analysis enables carrier detection and prenatal diagnosis. In a few centres, screening tests to identify the most common *CFTR* mutations are offered to pregnant women and their partners. If both partners carry an identifiable mutation, prenatal diagnosis can be offered prior to the birth of the first affected child.

Conventional treatment of CF involves pancreatic enzyme replacement and treatment of pulmonary infections with antibiotics and physiotherapy. These measures have dramatically improved survival rates for cystic fibrosis over the last 20 years. Several gene therapy trials have been undertaken in CF patients aimed at delivering the normal *CFTR* gene to the airway epithelium and research into this approach is continuing.

Cardiomyopathy

Several forms of cardiomyopathy are due to single gene defects, most being inherited in an autosomal dominant manner. The term cardiomyopathy was initially used to distinguish cardiac muscle disease of unknown origin from abnormalities secondary to hypertension, coronary artery disease and valvular disease.

Table 10.5 Frequency of cystic fibrosis mutations screened in the North-West of England

Mutation	Frequency (%)
G85E	0.3
R117H	0.7
621 + 1G→T	1.0
1078delT	0.1
ΔI507	0.5
ΔF508	88.0
1717-1G→T	0.3
G542X	1.3
S549N	0.2
G551D	4.2
R553X	0.7
R560T	0.7
1898 + 1G→A	1.0
3659delC	0.2
W1282X	0.3
N1303K	0.5

(Data provided by Dr M Schwarz M, Dr G M Malone, and Dr M Super, Central Manchester and Manchester Children's University Hospitals from 1254 CF chromosomes screened)

Box 10.6 Single gene disorders associated with congenital heart disease

- Holt Oram syndrome — Upper limb defects, atrial septal defect, cardiac conduction defect — autosomal dominant
- Noonan syndrome — 'Turner-like' phenotype, deafness, pulmonary stenosis, cardiomyopathy — autosomal dominant
- Leopard syndrome — multiple lentigenes, pulmonary stenosis, cardiac conduction defect — autosomal dominant
- Ellis-van Creveld — skeletal dysplasia, polydactyly, mid-line cleft lip — autosomal recessive
- Tuberous sclerosis — neurocutaneous features, hamartomas, cardiac leiomyomas — autosomal dominant

Table 10.6 Genes causing autosomal dominant hypertrophic obstructive cardiomyopathy

Gene product	Locus	Gene location
Cardiac myosin heavy chain α or β	FHC1	14q11.2
Cardiac troponin T	FHC2	1q32
Cardiac myosin binding protein C	FHC3	11p11.2
α Tropomyosin	FHC4	15q22
Regulatory myosin light chain	MYL2	12q23–q24
Essential myosin light chain	MYL3	3p21
Cardiac troponin I	TNNI3	19p12–q13
Cardiac alpha actin	ACTC	15q14

Hypertrophic cardiomyopathy (HOCM) has an incidence of about 1 in 1000. Presentation is with hypertrophy of the left and/or right ventricle without dilatation. Many affected individuals are asymptomatic and the initial presentation may be with sudden death. In others, there is slow progression of symptoms that include dyspnoea, chest pain and syncope. Myocardial hypertrophy may not be present before the adolescence growth spurt in children at risk, but a normal two-dimensional echocardiogram in young adults will virtually exclude the diagnosis. Many adults are asymptomatic and are diagnosed during family screening. Atrial or ventricular arrhythmias may be asymptomatic, but their presence indicates an increased likelihood of sudden death. Linkage analysis and positional cloning has identified several loci for HOCM. The genes known to be involved include those encoding for beta myosin heavy chain, cardiac troponin T, alpha tropomyosin and myosin binding protein C. These are sarcomeric proteins known to be essential for cardiac muscle contraction. Mutation analysis is not routine, but mutation detection allows presymptomatic predictive testing in family members at risk, identifying those relatives who require follow up.

Dilated cardiomyopathies demonstrate considerable heterogeneity. Autosomal dominant inheritance may account for about 25% of cases. Mutations in the cardiac alpha actin gene have been found in some autosomal dominant families and an X-linked form (Barth syndrome) is associated with skeletal myopathy, neutropenia and abnormal mitochondria due to mutations in the X-linked taffazin gene. Dystrophinopathy, caused by mutations in the X-linked gene causing Duchenne and Becker muscular dystrophies can sometimes present as isolated cardiomyopathy in the absence of skeletal muscle involvement.

Restrictive cardiomyopathy may be due to autosomal recessive inborn errors of metabolism that lead to accumulation of metabolites in the myocardium, to autosomal dominant familial amyloidosis or to autosomal dominant familial endocardial fibroelastosis.

Haematological disorders

Haemophilia
The term haemophilia has been used in reference to haemophilia A, haemophilia B and von Willebrand disease. Haemophilia A is the most common bleeding disorder affecting 1 in 5000 to 1 in 10 000 males. It is an X-linked recessive disorder due to deficiency of coagulation factor VIII. Clinical severity varies considerably and correlates with residual factor VIII activity. Activity of 1% leads to severe disease that occurs in about half of affected males and may present at birth. Activity of 1–5% leads to moderate disease, and 5–25% to mild disease that may not require treatment. Affected individuals have easy bruising, prolonged bleeding from wounds, and bleeding into muscles and joints after relatively mild trauma. Repeated bleeding into joints causes a chronic inflammatory reaction leading to haemophiliac arthropathy with loss of cartilage and reduced joint mobility. Treatment using human plasma or recombinant factor VIII controls acute episodes and is used electively for surgical procedures. Up to 15% of treated individuals develop neutralising antibodies that reduce the efficiency of treatment. Prior to 1984, haemophiliacs treated with blood products were exposed to the human immunodeficiency virus which resulted in a reduction in life expectancy to 49 years in 1990, compared to 70 years in 1980.

Table 10.7 Genetic disorders with associated cardiomyopathy

Condition	Inheritance
Duchenne and Becker muscular dystrophy	XLR
Emery–Dreifuss muscular dystrophy	XLR, AD
Mitochondrial myopathy	sporadic/maternal
Myotonic dystrophy	AD
Friedreich ataxia	AR
Noonan syndrome	AD

Box 10.7 Familial cardiac conduction defects

Long QT (Romano-Ward) syndrome
- autosomal dominant
- episodic dysrhythmias in a quarter of patients
- risk of sudden death
- several loci identified
- mutations found in sodium and potassium channel genes

Long QT (Jervell and Lange-Nielsen) syndrome
- autosomal recessive
- associated with congenital sensorineural deafness
- considerable risk of sudden death
- mutations found in potassium channel genes

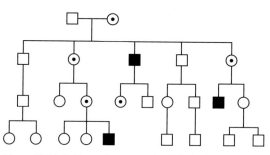

Figure 10.14 Pedigree demonstrating X linked recessive inheritance of Haemophilia A

Box 10.8 Haemochromatosis (HFE)

Common autosomal recessive disorder
- One in 10 of the population are heterozygotes
- Not all homozygotes are clinically affected

Clinical features
- Iron deposition can cause cirrhosis of the liver, diabetes, skin pigmentation and heart failure
- Primary hepatocellular carcinoma is responsible for one third of deaths in affected individuals

Management
- Early diagnosis and venesection prevents organ damage
- Normal life expectancy if venesection started in precirrhotic stage

Diagnosis
- Serum ferritin and fasting transferrin saturation levels
- Liver biopsy and hepatic iron index

Genetics
- Two common mutations in HFE gene: C282Y and H63D
- >80% of affected northern Europeans are homozygous for the C282Y mutation
- Role of H63D mutation (found in 20% of the population) less clear cut

The factor VIII gene (*F8C*) is located on the X chromosome at Xq28. Mutation analysis is used effectively in carrier detection and prenatal diagnosis. A range of mutations occur in the factor VIII gene with point mutations and inversion mutations predominating. The mutation rate in males is much greater than in females so that most mothers of isolated cases are carriers. This is because they are more likely to have inherited a mutation occurring during spermatogenesis transmitted by their father, than to have transmitted a new mutation arising during oogenesis to their sons.

Haemophilia B is less common than haemophilia A and also follows X-linked recessive inheritance, and is due to mutations in the factor IX gene (*F9*) located at Xq27. Mutations in this gene are usually point mutations or small deletions or duplications.

Renal disease

Adult polycystic kidney disease

Adult polycystic kidney disease (APKD) is typically a late onset, autosomal dominant disorder characterised by multiple renal cysts. It is one of the most common genetic diseases in humans and the incidence may be as high as 1 in 1000. There is considerable variation in the age at which end stage renal failure is reached and the frequency of hypertension, urinary tract infections, and hepatic cysts. Approximately 20% of APKD patients have end stage renal failure by the age of 50 and 70% by the age of 70, with 5% of all end stage renal failure being due to APKD. A high incidence of colonic diverticulae associated with a risk of colonic perforation is reported in APKD patients with end stage renal failure. An increased prevalence of 4–5% for intracranial aneurysms has been suggested, compared to the prevalence of 1% in the general population. There may also be an increased prevalence of mitral, aortic and tricuspid regurgitation, and tricuspid valve prolapse in APKD.

All affected individuals have renal cysts detectable on ultrasound scan by the age of 30. Screening young adults at risk will identify those asymptomatic individuals who are affected and require annual screening for hypertension, urinary tract infections and decreased renal function. Children diagnosed under the age of one year may have deterioration of renal function during childhood, but there is little evidence that early detection in asymptomatic children affects prognosis.

There is locus heterogeneity in APKD with at least three loci identified by linkage studies and two genes cloned. The gene for APKD1 on chromosome 16p encodes a protein called polycystin-1, which is an integral membrane protein involved in cell–cell/matrix interactions. The protein encoded by the gene for APKD2 on chromosome 4 has been called polycystin-2. Mutation analysis is not routinely undertaken, but linkage studies may be used in conjunction with ultrasound scanning to detect asymptomatic gene carriers.

Deafness

Severe congenital deafness

Severe congenital deafness affects approximately 1 in 1000 infants. This may occur as an isolated deafness as or part of a syndrome. At least half the cases of congenital deafness have a genetic aetiology. Of genetic cases, approximately 66% are autosomal recessive, 31% are autosomal dominant, 3% are X linked recessive. Over 30 autosomal recessive loci have been identified. This means that two parents with autosomal recessive congenital deafness will have no deaf children if their

Table 10.8 Examples of single gene disorder with renal manifestations

Disorder	Features	Inheritance
Tuberous sclerosis	Multiple hamartomas Epilepsy Intellectual retardation Renal cysts/angiomyolipomas	AD
von Hippel-Lindau disease	Retinal angiomas Cerebellar haemangioblastomas Renal cell carcinoma	AD
Infantile polycystic kidney disease	Renal and hepatic cysts (histological diagnosis required)	AR
Cystinuria	Increased dibasic amino acid excretion Renal calculi	AR
Cystinosis	Cystine storage disorder Progressive renal failure	AR
Jeune syndrome	Thoracic dysplasia Renal dysplasia	AR
Meckel syndrome	Encephalocele Polydactyly Renal cysts	AR
Alport syndrome	Deafness Microscopic haematuria Renal failure	X-linked/AD
Fabry disease	Skin lesions Cardiac involvement Renal failure	XLR
Lesch–Nyhan syndrome	Intellectual retardation Athetosis Self-mutilation Uric acid stones	XLR
Lowe syndrome	Intellectual retardation Cataracts Renal tubular acidosis	XLR

Table 10.9 Examples of syndromes associated with deafness

Condition	Features	Inheritance
Pendred syndrome	Severe nerve deafness Thyroid goitre	AR
Usher syndrome	Nerve deafness Retinitis pigmentosa	AR
Jervell–Lange–Nielson syndrome	Nerve deafness Cardiac conduction defect	AR
Treacher Collins syndrome	Nerve deafness Mandibulo-facial dysostosis	AD
Waardenberg syndrome	Nerve deafness Pigmentary abnormalities	AD
Branchio-otorenal syndrome	Nerve deafness Branchial cysts Renal anomalies	AD
Stickler syndrome	Nerve deafness Myopia Cleft palate Arthropathy	AD
Alport syndrome	Nerve deafness Microscopic haematuria Renal failure	X linked/AD

own deafness is due to different genes, but all deaf children if the same gene is involved.

Connexin 26 mutations

Mutations in the connexin 26 gene (*CX26*) on chromosome 13 have been found in severe autosomal recessive congenital deafness and may account for up to 50% of cases. One specific mutation, 30delG accounts for over half of the mutations detected. The carrier frequency for *CX26* mutations in the general population is around 1 in 35. Mutation analysis in affected children enables carrier detection in relatives, early diagnosis in subsequent siblings and prenatal diagnosis if requested.

The *CX26* gene encodes a gap junction protein that forms plasma membrane channels that allow small molecules and ions to move from one cell to another. These channels play a role in potassium homeostasis in the cochlea which is important for inner ear function.

Pendred syndrome

Pendred syndrome is an autosomal recessive form of deafness due to cochlear abnormality that is associated with a thyroid goitre. It may account for up to 10% of hereditary deafness. Not all patients have thyroid involvement at the time the deafness is diagnosed and the perchlorate discharge test has been used in diagnosis.

The gene for Pendred syndrome, called *PDS*, was isolated in 1997 and is located on chromosome 7. The protein product called pendrin, is closely related to a number of sulphate transporters and is expressed in the thyroid gland. Mutation detection enables diagnosis and carrier testing within affected families.

Eye disorders

Both childhood onset severe visual handicap and later onset progressive blindness commonly have a genetic aetiology. X linked inheritance is common, but there are also many autosomal dominant and recessive conditions. Leber hereditary optic neuropathy is a late onset disorder causing rapid development of blindness that follows maternal inheritance from an underlying mitochondrial DNA mutation. Genes for a considerable number of a mendelian eye disorders have been identified. Mutation analysis will increasingly contribute to clinical diagnosis since the mode of inheritance can often not be determined from clinical presentation in sporadic cases. Mutation analysis will also be particularly useful for carrier detection in females with a family history of X linked blindness.

Retinitis pigmentosa

Retinitis pigmentosa (RP) is the most common type of inherited retinal degenerative disorder. Like many other eye conditions it is genetically heterogeneous, with autosomal dominant (25%), autosomal recessive (50%), and X linked (25%) cases. In isolated cases the mode of inheritance cannot be determined from clinical findings, except that X linked inheritance can be identified if female relatives have pigmentary abnormalities and an abnormal electroretinogram. Linkage studies have identified three gene loci for X linked retinitis pigmentosa and mutations in the rhodopsin and peripherin genes occur in a significant proportion of dominant cases.

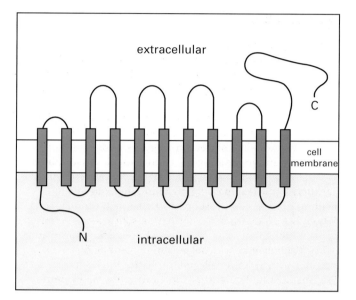

Figure 10.15 Diagramatic representation of the pendrin protein which has intracellular, extracellular and transmembrane domains. Mutations in each of these domains have been identified in the pendrin protein gene in different people with Pendred syndrome

---

Box 10.9 Examples of autosomal dominant eye disorders

- Late onset macular dystrophies
- Best macular degeneration
- Retinitis pigmentosa (some types)
- Hereditary optic atrophy (some types)
- Corneal dystrophies (some types)
- Stickler syndrome (retinal detachment)
- Congenital cataracts (some types)
- Lens dislocation (Marfan syndrome)
- Hereditary ptosis
- Microphthalmia with coloboma

Box 10.10 Examples of autosomal recessive eye disorders

- Juvenile Stargardt macular dystrophy
- Retinitis pigmentosa (some types)
- Leber congenital amaurosis
- Hereditary optic atrophy (some types)
- Congenital cataracts (some types)
- Lens dislocation (homocystinuria)
- Congenital glaucoma (some types)
- Complete bilateral anophthalmia

Box 10.11 Examples of X-linked recessive eye disorders

- Colour blindness
- Ocular albinisim
- Hereditary oculomotor nystagmus
- Choroideraemia
- Retinoschisis
- Lenz microphthalmia syndrome
- Norrie disease (pseudoglioma)
- Lowe oculocerebrorenal syndrome
- X linked retinitis pigmentosa
- X linked congenital cataract
- X linked macular dystrophy

Skin diseases

Epidermolysis bullosa

Epidermolysis bullosa (EB) is a clinically and genetically heterogeneous group of blistering skin diseases. The main types are designated as simplex, junctional and dystrophic, based on ultrastructural analysis of skin biopsies. EB simplex causes recurrent, non-scarring blisters from increased skin fragility. The majority of cases are due to autosomal dominant mutations in either the keratin 5 or keratin 14 genes. A rare autosomal recessive syndrome of EB simplex and muscular dystrophy is due to a mutation in a gene encoding plectin. Junctional EB is characterised by extreme fragility of the skin and mucus membranes with blisters occurring after minor trauma or friction. Both lethal and non-lethal autosomal recessive forms occur and mutations have been found in several genes that encode basal lamina proteins, including laminin 5, integrin and type XVII collagen. In dystrophic EB the blisters cause mutilating scars and gastrointestinal strictures, and there is an increased risk of severe squamous cell carcinomas in affected individuals. Autosomal recessive and dominant cases caused by mutations in the collagen VII gene.

Mutation analysis in specialist centres enables prenatal diagnosis in families, which is particularly appropriate for the more severe forms of the disease. Skin disorders such as epidermolysis bullosa provide potential candidates for gene therapy, since the affected tissue is easily accessible and amenable to a variety of potential in vivo and ex vivo gene therapy approaches.

Table 10.10 Examples of mendelian disorders affecting the epidermis

Condition	Inheritance
Ectodermal dysplasias	
Ectrodactyly/ectodermal dysplasia/clefting	AD
Rapp–Hodgkin ectodermal dysplasia	AD
Hypohydrotic ectodermal dysplasia	AR/XLR
Goltz focal dermal hypoplasia	XLD
Incontinentia pigmenti	XLD
Ichthyoses	
Ichthyosis vulgaris	AD
Steroid sulphatase deficiency	XLR
Lamellar ichthyosis	AD/AR
Bullous ichthyosiform erythroderma	AD
Non-bullous ichthyosiform erythroderma	AR
Sjögren–Larsson syndrome	AR
Refsum syndrome	AR
Keratodermas	
Vohwinkel mutilating	AD
Pachyonychia congenita	AD
Papillon le Fevre	AR
Palmoplantar keratoderma with leucoplakia	AD
Follicular hyperkeratoses	
Darrier disease	AD

11 Genetics of cancer

Cellular proliferation is under genetic control and development of cancer is related to a combination of environmental mutagens, somatic mutation and inherited predisposition. Molecular studies have shown that several mutational events, that enhance cell proliferation and increase genome instability, are required for the development of malignancy. In familial cancers one of these mutations is inherited and represents a constitutional change in all cells, increasing the likelihood of further somatic mutations occurring in the cells that lead to tumour formation. Chromosomal translocations have been recognised for many years as being markers for, or the cause of, certain neoplasms, and various oncogenes have been implicated.

The risk that a common cancer will occur in relatives of an affected person is generally low, but familial aggregations that cannot be explained by environmental factors alone exist for some neoplasms. Up to 5% of cases of breast, ovary, and bowel cancers are inherited because of mutations in incompletely penetrant, autosomal dominant genes. There are also several cancer predisposing syndromes that are inherited in a mendelian fashion, and the genes responsible for many of these have been cloned.

Mechanisms of tumorigenesis

The genetic basis of both sporadic and inherited cancers has been confirmed by molecular studies. The three main classes of genes known to predispose to malignancy are oncogenes, tumour suppressor genes and genes involved in DNA mismatch repair. In addition, specific mutagenic defects from environmental carcinogens and viral infections (notably hepatitis B) have been identified.

Oncogenes are genes that can cause malignant transformation of normal cells. They were first recognised as viral oncogenes (v-onc) carried by RNA viruses. These retroviruses incorporate a DNA copy of their genomic RNA into host DNA and cause neoplasia in animals. Sequences homologous to those of viral oncogenes were subsequently detected in the human genome and called cellular oncogenes (c-onc). Numerous proto-oncogenes have now been identified, whose normal function is to promote cell growth and differentiation. Mutation in a proto-oncogene results in altered, enhanced, or inappropriate expression of the gene product leading to neoplasia. Oncogenes act in a dominant fashion in tumour cells, i.e. mutation in one copy of the gene is sufficient to cause neoplasia. Proto-oncogenes may be activated by point mutations, but also by mutations that do not alter the coding sequence, such as gene amplification or chromosomal translocation. Most proto-oncogene mutations occur at a somatic level, causing sporadic cancers. Exceptions include the germline mutation in the *RET* oncogene responsible for dominantly inherited multiple endocrine neoplasia type II.

Tumour suppressor genes normally act to inhibit cell proliferation by stopping cell division, initiating apoptosis (cell death) or being involved in DNA repair mechanisms. Loss of function or inactivation of these genes is associated with tumorigenesis. At the cellular level these genes act in a recessive fashion, as loss of activity of both copies of the gene is required for malignancy to develop. Mutations inactivating various tumour suppressor genes are found in both sporadic and hereditary cancers.

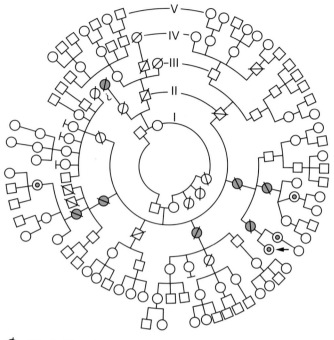

Φ Affected females

⊙ Females at up to 50% risk having undergone prophylatic oophorectomy

Figure 11.1 Autosomal dominant inheritance of ovarian cancer (courtesy of Professor Dian Donnai, Regional Genetic Service, St Mary's Hospital, Manchester)

Table 11.1 Cloned genes in dominantly inherited cancers

Cancers	Gene symbol	Gene type*	Chromosomal localisation
Familial common cancers			
Familial adenomatous polyposis	APC	TS	5q21
HNPCC	hMSH2	Mis	2p16
	hMLH1	Mis	3p21.3-23
	hPMS1	Mis	2q31-33
	hPMS2	Mis	7p22
	MSH6	Mis	2p16
Familial breast–ovarian cancer	BRAC1	TS	17q21
	BRAC2	TS	13q12-13
Li–Fraumeni syndrome	TP53	TS	17p13
Familial melanoma	MLM	TS	9q21
Cancer syndromes			
Basal cell naevus syndrome	PTCH	TS	9q31
Multiple endocrine neoplasia 1	MEN1	TS	11q13
Multiple endocrine neoplasia 2	RET	Onc	10q11
Neurofibromatosis type 1	NFI	TS	17q11
Neurofibromatosis type 2	NF2	TS	22q12
Retinoblastoma	RB1	TS	13q14
Tuberous sclerosis 1	TSC1	TS	9q34
Tuberous sclerosis 2	TSC2	TS	16p13
von Hippel–Lindau disease	VHL	TS	3p25
Renal cell carcinoma	MET	Onc	7q31
Wilms tumour	WT1	TS	11p13
Tylosis	TOC	TS	17q24

*TS=tumour suppressor; Onc=oncogene; Mis=mismatch repair

Another mechanism for tumour development is the failure to repair damaged DNA. Xeroderma pigmentosum, for example, is a rare autosomal recessive disorder caused by failure to repair DNA damaged by ultraviolet light. Exposure to sunlight causes multiple skin tumours in affected individuals. Many other tumours are found to be associated with instability of multiple microsatellite markers because of a failure to repair mutated DNA containing mismatched base pairs. Microsatellite instability is particularly common in colorectal, gastric and endometrial cancers. Hereditary non-polyposis colon cancer (HNPCC) is due to mutations in genes on chromosomes 2p, 2q, 3p and 7p. The *hMSH2* gene on chromosome 2p represents a mismatch repair gene. Some patients with HNPCC inherit one mutant copy of this gene, which is inactivated in all cells. Loss of the other allele (loss of heterozygosity) in colonic cells leads to an increase in the mutation rate in other genes, resulting in the development of colonic cancer.

The most commonly altered gene in human cancers is the tumour suppressor gene *TP53* which encodes the p53 protein. *TP53* mutations are found in about 70% of all tumours. Mutations in the *RAS* oncogene occur in about one third. Interestingly, somatic mutations in the tumour suppressor gene *TP53* are often found in sporadic carcinoma of the colon, but germline mutation of *TP53* (responsible for Li–Fraumeni syndrome) seldom predisposes to colonic cancer. Similarly, lung cancers often show somatic mutations of the retinoblastoma (*RB1*) gene, but this tumour does not occur in individuals who inherit germline *RB1* mutations. These genes probably play a greater role in progression, than in initiation, of these tumours. Although caused by mutations in the *hMSH2* gene, the colonic cancers commonly associated with HNPCC show somatic mutations similar to those found in sporadic colon cancers, that is in the adenomatous polyposis coli (*APC*) gene, *K-RAS* oncogene and *TP53* tumour suppressor. This is because the HNPCC predisposing mismatch repair genes are acting as mutagenic rather than tumour suppressor genes.

There now exists the possibility of gene therapy for cancers, and many of the protocols currently approved for genetic therapy are for patients with cancer. Several approaches are being investigated, including virally directed enzyme prodrug therapy, the use of transduced tumour infiltrating lymphocytes, which produce toxic gene products, modifying tumour immunogenicity by inserting genes, or the direct manipulation of crucial oncogenes or tumour suppressor genes.

Chromosomal abnormalities in malignancy

Structural chromosomal abnormalities are well documented in leukaemias and lymphomas and are used as prognostic indicators. They are also evident in solid tumours, for example, an interstitial deletion of chromosome 3 occurs in small cell carcinoma of the lung. More than 100 chromosomal translocations are associated with carcinogenesis, which in many cases is caused by ectopic expression of chimaeric fusion proteins in inappropriate cell types. In addition, chromosome instability is seen in some autosomal recessive disorders that predispose to malignancy, such as ataxia telangiectasia, Fanconi anaemia, xeroderma pigmentosum, and Bloom syndrome.

Philadelphia chromosome
The Philadelphia chromosome, found in blood and bone marrow cells, is a deleted chromosome 22 in which the long arm has been translocated on to the long arm of chromosome 9 and is designated t(9;22) (q34;ql, 1).

Table 11.2 Examples of proto-oncogenes implicated in human malignancy

Proto-oncogene	Molecular abnormality	Disorder
myc	Translocation 8q24	Burkitt lymphoma
abl	Translocation 9q34	Chronic myeloid leukaemia
mos	Translocation 8q22	Acute myeloid leukaemia
myc	Amplification	Carcinoma of breast, lung, cervix, oesophagus
N-myc	Amplification	Neuroblastoma, small cell carcinoma of lung
K-ras	Point mutation	Carcinoma of colon, lung and pancreas; melanoma
H-ras	Point mutation	Carcinoma of genitourinary tract, thyroid

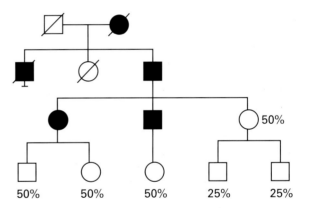

Figure 11.2 Family with autosomal dominant hereditary non-polyposis colon cancer (HNPCC) indicating individuals at risk who require investigation (■ ● affected individuals)

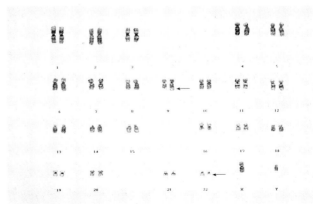

Figure 11.3 9; 22 translocation in chronic myeloid leukaemia producing the Philadelphia chromosome (deleted chromosome 22) (courtesy of Oncology Cytogenetic Services, Christie Hospital, Manchester)

The translocation occurs in 90% of patients with chronic myeloid leukaemia, and its absence generally indicates a poor prognosis. The Philadelphia chromosome is also found in 10–15% of acute lymphocytic leukaemias, when its presence indicates a poor prognosis.

Burkitt lymphoma

Burkitt lymphoma is common in children in parts of tropical Africa. Infection with Epstein–Barr (EB) virus and chronic antigenic stimulation with malaria both play a part in the pathogenesis of the tumour. Most lymphoma cells carry an 8;14 translocation or occasionally a 2;8 or 8;22 translocation. The break points involve the *MYC* oncogene on chromosome 8 at 8q24, the immunoglobulin heavy chain gene on chromosome 14, and the *K* and *A* light chain genes on chromosomes 2 and 22 respectively. Altered activity of the oncogene when translocated into regions of immunoglobulin genes that are normally undergoing considerable recombination and mutation plays an important part in the development of the tumour.

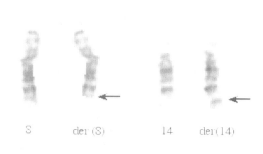

Figure 11.4 8;14 translocation in Burkitt lymphoma (courtesy of Oncology Cytogenetics Service, Christie Hospital, Manchester)

Inherited forms of common cancers

Inherited forms of the common cancers, notably breast, ovary and bowel, constitute a small proportion of all cases, but their identification is important because of the high risk of malignancy associated with inherited mutations in cancer predisposing genes. Identification of such families can be difficult, as tumours often vary in the site of origin, and the risk and type of malignancy may vary with sex. For example, in HNPCC, females have a higher risk of uterine cancer than bowel cancer. In breast or breast–ovary cancer families, most males carrying the predisposing mutations will manifest no signs of doing so, but their daughters will be at 50% risk of inheriting a mutation, associated with an 80% risk of developing breast cancer. With the exception of familial adenomatosis polyposis (FAP, see below), where the sheer number of polyps or systemic manifestations may lead to the correct diagnosis, pathological examination of most common tumours does not usually help in determining whether or not a particular malignancy is due to an inherited gene mutation, since morphological changes are seldom specific or invariable. Determining the probability that any particular malignancy is inherited requires an accurate analysis of a three-generation family tree. Factors of importance are the number of people with a malignancy on both maternal and paternal sides of the family, the types of cancer that have occurred, the relationship of affected people to each other, the age at which the cancer occurred, and whether or not a family member has developed two or more cancers. A positive family history becomes more significant in ethnic groups where a particular cancer is rare. In other ethnic groups there may be a particularly high population incidence of particular mutations, such as the *BRCA1* and *BRCA2* mutations occurring in people of Jewish Ashkenazi origin.

Epidemiological studies suggest that mutations in *BRCA1* account for 2% of all breast cancers and, at most, 5% of ovarian cancer. Mutations in *BRCA2* account for less than 2% of breast cancer in women, 10% of breast cancer in men and 1% of ovarian cancer. Most clustering of breast cancer in families is therefore probably due to the influence of other, as yet unidentified, genes of lower penetrance, with or without an effect from modifying environmental factors.

Box 11.1 Types of tumour in inherited cancer families

BRCA1
- Breast, ovary
- Prostate, bowel (lower risk)

BRCA2
- Breast, ovary
- Stomach, pancreas, prostate, thyroid,
- Hodgkin lymphoma, gallbladder (risk lower and influenced by mutation)

HNPCC
- Colon
- Endometrium
- Upper ureter or renal pelvis
- Ovary
- Stomach, oesophagus, small bowel
- Pancreas, biliary tree, larynx

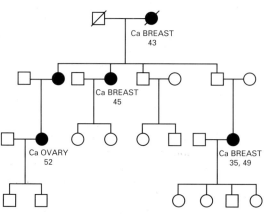

Figure 11.5 Pedigree demonstrating autosomal dominant inheritance of a *BRCA1* mutation with transmission of the mutant gene through an unaffected male to his daughter

Hereditary non-polyposis colon cancer (HNPCC) has been called Lynch syndrome type I in families where only bowel cancer is present, and Lynch syndrome type II in families with bowel cancer and other malignancies. HNPCC is due to inheritance of autosomal genes that act in a dominant fashion and accounts for 1–2% of all bowel cancer. In most cases of bowel cancer, a contribution from other genes of moderate penetrance, with or without genetic modifiers and environmental triggers seems the likely cause.

Gene testing to confirm a high genetic risk of malignancy has received a lot of publicity, but is useful in the minority of people with a family history, and requires identification of the mutation in an affected person as a prerequisite. When the family history clearly indicates an autosomal dominant pattern of inheritance, risk determination is based on a person's position in the pedigree and the risk and type of malignancy associated with the mutation. In families where an autosomal dominant mode of transmission appears unlikely, risk is determined from empiric data. Studies of large numbers of families with cancer have provided information as to how likely a cancer predisposing mutation is for a given family pedigree. These probabilities are reflected in guidelines for referral to regional genetic services.

Management of those at increased risk of malignancy because of a family history is based on screening. Annual mammography between ages 35 and 50 is suggested for women at 1 in 6 or greater risk of breast cancer, and annual transvaginal ultrasound for those at 1 in 10 or greater risk of ovarian cancer. In HNPCC (as in the general population), all bowel malignancy arises in adenomatous polyps, and regular colonoscopy with removal of polyps is offered to people whose risk of bowel cancer is 1 in 10 or greater. The screening interval and any other screening tests needed are influenced by both the pedigree and tumour characteristics.

Table 11.3 Guidelines for referral to a regional genetics service

Breast cancer*
- Four or more relatives diagnosed at any age
- Three close relatives diagnosed less than 60
- Two close relatives diagnosed under 50
- Mother or sister diagnosed under 40
- Father or brother with breast cancer diagnosed at any age
- One close relative with bilateral breast cancer diagnosed at any age

Ovarian cancer and breast/ovarian cancer*
- Three or more close relatives diagnosed with ovarian cancer at any age
- Two close relatives diagnosed with ovarian cancer under 60
- One close relative diagnosed with ovarian cancer at any age and at least two close relatives diagnosed with breast cancer under 60
- One close relative diagnosed with ovarian cancer at any age and at least 1 close relative diagnosed with breast cancer under 50
- One close relative diagnosed with breast and ovarian cancer at any age

A close relative means a parent, brother, sister, child, grandparent, aunt, uncle, nephew or niece.

*Cancer Research Campaign Primary Care Education Research Group

Bowel cancer[+]
- One close relative diagnosed less than 35 years
- Two close relatives with average age of diagnosis less than 60 years
- Three or more relatives with bowel cancer on the same side of the family
- Bowel and endometrial cancer in the same person, with diagnosis less than 50 years

+North West Regional Genetic Service, suggested guidelines

Inherited cancer syndromes

Multiple polyposis syndromes

Familial adenomatous polyposis (FAP) follows autosomal dominant inheritance and carries a high risk of malignancy necessitating prophylactic colectomy. The presentation may be with adenomatous polyposis as the only feature or as the Gardener phenotype in which there are extracolonic manifestations including osteomas, epidermoid cysts, upper gastrointestinal polyps and adenocarcinomas (especially duodenal), and desmoid tumours that are often retroperitoneal. Detecting congenital hypertrophy of the retinal pigment epithelium (CHRPE), that occurs in familial adenomatous polyposis, has been used as a method of early identification of gene carriers. The adenomatous polyposis coli (*APC*) gene on chromosome 5 responsible for FAP has been cloned. Mutation detection or linkage analysis in affected families provides a predictive test to identify gene carriers. Family members at risk should be screened with regular colonoscopy from the age of 10 years.

In Peutz–Jeghers syndrome hamartomatous gastrointestinal polyps, which may bleed or cause intussusception, are associated with pigmentation of the buccal mucosa and lips. Malignant degeneration in the polyps occurs in up to 30–40% of cases. Ovarian, breast and endometrial tumours also occur in this dominant syndrome.

Mutations causing Peutz–Jeghers syndrome have been detected in the serine/threonine protein kinase gene (*STK11*) on chromosome 19p13.3.

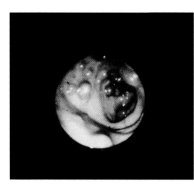

Figure 11.6 Colonic polyps in familial adenomatous polyposis (courtesy of Gower Medical Publishing and Dr C Williams, St Mary's Hospital, London)

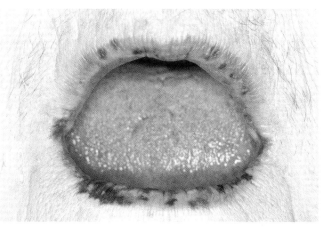

Figure 11.7 Pigmentation of lips in Peutz-Jehger syndrome

Li–Fraumeni syndrome

Li–Fraumeni syndrome is a dominantly inherited cancer syndrome caused by constitutional mutations in the *TP53* or *CHK2* genes. Affected family members develop multiple primary tumours at an early age that include rhabdomyosarcomas, soft tissue sarcomas, breast cancer, brain tumours, osteosarcomas, leukaemia, adrenocortical carcinoma, lymphomas, lung adenocarcinoma, melanoma, gonadal germ cell tumours, prostate carcinoma and pancreatic carcinoma. Mutation analysis may confirm the diagnosis in a family and enable predictive genetic testing of relatives, but screening for neoplastic disease in those at risk is difficult.

Multiple endocrine neoplasia syndromes

Two main types of multiple endocrine neoplasia syndrome exist and both follow autosomal dominant inheritance with reduced penetrance. Many affected people have involvement of more than one gland but the type of tumour and age at which these develop is very variable within families. The gene for MEN type I on chromosome 11 acts as a tumour suppressor gene and encodes a protein called menin. Mutations in the coding region of the gene are found in 90% of individuals with a diagnosis of MEN I based on clinical criteria. First-degree relatives in affected families should be offered predictive genetic testing. Those carrying the mutation require clinical, biochemical and radiological screening to detect presymptomatic tumours. MEN type II is due to mutations in the *RET* oncogene on chromosome 10 that encodes a tyrosine kinase receptor protein. Mutation analysis again provides confirmation of the diagnosis in the index case and presymptomatic tests for relatives. Screening tests in gene carriers include calcium or pentagastrin provocation tests that detect abnormal calcitonin secretion and permit curative thyroidectomy before the tumour cells extend beyond the thyroid capsule.

von Hippel–Lindau disease

In von Hippel–Lindau disease haemangioblastomas develop throughout the brain and spinal cord, characteristically affecting the cerebellum and retina. Renal, hepatic and pancreatic cysts also occur. The risk of clear cell carcinoma of the kidney is high and increases with age. Phaeochromocytomas occur but are less common. The syndrome follows autosomal dominant inheritance, and clinical, biochemical and radiological screening is recommended for affected family members and those at risk, to permit early treatment of problems as they arise. The *VHL* gene on chromosome 3 has been cloned, and identification of mutations allows predictive testing in the majority of families.

Naevoid basal cell carcinoma

The cardinal features of the naevoid basal cell carcinoma syndrome, an autosomal dominant disorder delineated by Gorlin, are basal cell carcinomas, jaw cysts and various skeletal abnormalities, including bifid ribs. Other features are macrocephaly, tall stature, palmar pits, calcification of the falx cerebri, ovarian fibromas, medulloblastomas and other tumours. The skin tumours may be extremely numerous and are usually bilateral and symmetrical, appearing over the face, neck, trunk, and arms during childhood or adolescence. Malignant change is very common after the second decade, and removal of the tumours is therefore indicated. Medulloblastomas occur in about 5% of cases. Abnormal sensitivity to therapeutic doses of ionising radiation results in the development of multiple basal cell carcinomas in any irradiated area. The gene for Gorlin syndrome (*PTCH*) on chromosome 9 has been cloned and is homologous to a drosophila developmental gene called *patched*.

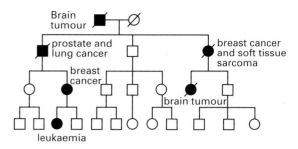

Figure 11.8 Multiple malignancies occuring at a young age in a family with Li–Fraumeni syndrome caused by a mutation in the *TP53* gene

Table 11.4 Main types of multiple endocrine neoplasia

MEN type I	MEN type II
Parathyroid 95%	Medullary thyroid MEN 99%
Pancreatic islet 40%	Phaeochromocytoma 50%
Anterior pituitary 30%	Parathroid 20%
Associated tumours:	
carcinoid, adrenocortical carcinoma, lipomas, angiofibromas, collagenomas	mucosal neuromas

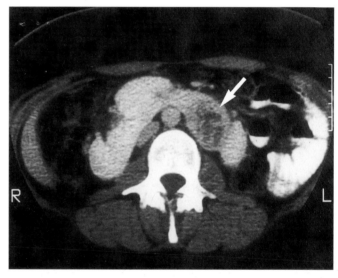

Figure 11.9 Renal carcinoma in horsehoe kidney on abdominal CT scan in von Hippel–Lindau disease

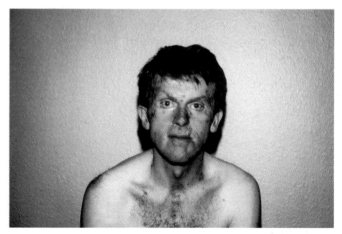

Figure 11.10 Multiple basal cell carcinomas in Gorlin syndrome (courtesy of Professor Gareth Evans, Regional Genetic Service, St Mary's Hospital, Manchester)

Neurofibromatosis

The presenting features of neurofibromatosis type 1 (NF1, peripheral neurofibromatosis, von Recklinghausen disease) and neurofibromatosis type 2 (NF2, central neurofibromatosis) are described in chapter 10. Benign optic gliomas and spinal neurofibromas may occur in NF1 and malignant tumours, mainly neurofibrosarcomas or embryonal tumours, occur in 5% of affected people. The gene for NF1 on chromosome 17 has been cloned, but mutation analysis is not routinely undertaken because of the large size of the gene (60 exons) and the diversity of mutations occurring. Deletions of the entire gene have been found in more severely affected cases.

The main feature of NF2 is bilateral acoustic neuromas (vestibular schwannomas). Spinal tumours and intracranial meningiomas occur in over 40% of cases. Surgical removal of VIIIth nerve tumours is difficult and prognosis for this disorder is often poor. The *NF2* gene on chromosome 22 has been cloned and various mutations, deletions and translocations have been identified, allowing presymptomatic screening and prenatal diagnosis within affected families.

Tuberous sclerosis

Tuberous sclerosis is an autosomal dominant disorder, very variable in its manifestation, that can cause epilepsy and severe retardation in affected children. Hamartomas of the brain, heart, kidney, retina and skin may also occur, and their presence indicates the carrier state in otherwise healthy family members. Sarcomatous malignant change is possible but uncommon. Tuberous sclerosis can be due to mutations in genes on chromosomes 9 and 16 (*TSC1* and *TSC2*).

Childhood tumours

Retinoblastoma

Sixty percent of retinoblastomas are sporadic and unilateral, with 40% being hereditary and usually bilateral. Hereditary retinoblastomas follow an autosomal dominant pattern of inheritance with incomplete penetrance. About 80–90% of children inheriting the abnormal gene will develop retinoblastomas. Molecular studies indicate that two events are involved in the development of the tumour, consistent with Knudson's original "two hit" hypothesis. In bilateral tumours the first mutation is inherited and the second is a somatic event with a likelihood of occurrence of almost 100% in retinal cells. In unilateral tumours both events probably represent new somatic mutations. The retinoblastoma gene is therefore acting recessively as a tumour suppressor gene.

Tumours may occasionally regress spontaneously leaving retinal scars, and parents of an affected child should be examined carefully. Second malignancies occur in up to 15% of survivors in familial cases. In addition to tumours of the head and neck caused by local irradiation treatment, other associated malignancies include sarcomas (particularly of the femur), breast cancers, pinealomas and bladder carcinomas.

A deletion on chromosome 13 found in a group of affected children, some of whom had additional congenital abnormalities, enabled localisation of the retinoblastoma gene to chromosome 13q14. The esterase D locus is closely linked to the retinoblastoma locus and was used initially as a marker to identify gene carriers in affected families. The retinoblastoma gene has now been cloned and mutation analysis is possible.

Wilms tumour

Wilms tumours are one of the most common solid tumours of childhood, affecting 1 in 10 000 children. Wilms tumours are

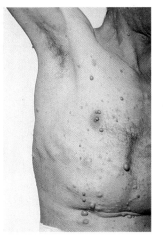

Figure 11.11 Neurofibromatosis type 1

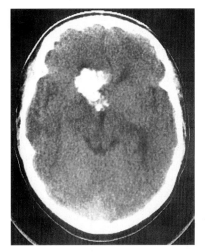

Figure 11.12 Heavily calcified intracranial hamartoma in tuberous sclerosis

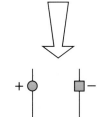
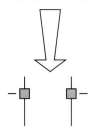

Inherited mutation

Chromosome rearrangement with gene disruption

New gene deletion or point mutation

First event

+ Normal allele
– Mutant allele

Loss of normal chromosome and duplication of abnormal chromosome

Recombination between chromosomes in mitosis

New gene deletion or point mutation

Second event

Figure 11.13 Two stages of tumour generation

usually unilateral, and the vast majority are sporadic. About 1% of Wilms tumours are hereditary, and of these about 20% are bilateral. Wilms tumour is associated with aniridia, genitourinary abnormalities, gonadoblastoma and mental retardation (WAGR syndrome) in a small proportion of cases. Identification of an interstitial deletion of chromosome 11 in such cases localised a susceptibility gene to chromosome 11p13. The Wilms tumour gene, *WT1*, at this locus has now been cloned and acts as a tumour suppressor gene, with loss of alleles on both chromosomes being detected in tumour tissue. A second locus at 11p15 has also been implicated in Wilms tumour. The insulin-like growth factor-2 gene (*IGF2*), is located at 11p15 and causes Beckwith–Wiedemann syndrome, an overgrowth syndrome predisposing to Wilms tumour. Children with hemihypertrophy are at increased risk of developing Wilms tumours and a recommendation has been made that they should be screened using ultrasound scans and abdominal palpation during childhood. A third gene predisposing to Wilms tumour has been localised to chromosome 16q. These genes are not implicated in familial Wilms tumour, which follows autosomal dominant inheritance with reduced penetrance, and there is evidence for localisation of a familial predisposition gene at chromosome 17q.

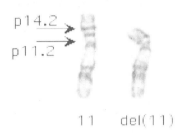

Figure 11.14 Deletion of chromosome 11 including band 11p13 is associated with Wilms tumour (courtesy of Dr Lorraine Gaunt and Helena Elliott, Regional Genetic Service, St Mary's Hospital, Manchester)

12 Genetics of common disorders

The genetic contribution to disease varies; some disorders are entirely environmental and others are wholly genetic. Many common disorders, however, have an appreciable genetic contribution but do not follow simple patterns of inheritance within a family. The terms multifactorial or polygenic inheritance have been used to describe the aetiology of these disorders. The positional cloning of multifactorial disease genes presents a major challenge in human genetics.

Multifactorial inheritance

The concept of multifactorial inheritance implies that a disease is caused by the interaction of several adverse genetic and environmental factors. The liability of a population to a particular disease follows a normal distribution curve, most people showing only moderate susceptibility and remaining unaffected. Only when a certain threshold of liability is exceeded is the disorder manifest. Relatives of an affected person will show a shift in liability, with a greater proportion of them being beyond the threshold. Familial clustering of a particular disorder may therefore occur. Genetic susceptibility to common disorders is likely to be due to sequence variation in a number of genes, each of which has a small effect, unlike the pathogenic mutations seen in mendelian disorders. These variations will also be seen in the general population and it is only in combination with other genetic variations that disease susceptibility becomes manifest.

Unravelling the molecular genetics of the complex multifactorial diseases is much more difficult than for single gene disorders. Nevertheless, this is an important task as these diseases account for the great majority of morbidity and mortality in developed countries. Approaches to multifactorial disorders include the identification of disease associations in the general population, linkage analysis in affected families, and the study of animal models. Identification of genes causing the familial cases of diseases that are usually sporadic, such as Alzheimer disease and motor neurone disease, may give insights into the pathogenesis of the more common sporadic forms of the disease. In the future, understanding genetic susceptibility may enable screening for, and prevention of, common diseases as well as identifying people likely to respond to particular drug regimes.

Several common disorders thought to follow polygenic inheritance (such as diabetes, hypertension, congenital heart disease and Hirschsprung disease) have been found in some individuals and families to be due to single gene defects. In Hirschsprung disease (aganglionic megacolon) family data on recurrence risks support the concept of sex-modified polygenic inheritance, although autosomal dominant inheritance with reduced penetrance has been suggested in some families with several affected members. Mutations in the ret proto-oncogene on chromosome 10q11.2 or in the endothelin-B receptor gene on chromosome 13q22 have been detected in both familial and sporadic cases, indicating that a proportion of cases are due to a single gene defect.

Risk of recurrence
The risk of recurrence for a multifactorial disorder within a family is generally low and mainly affects first degree relatives. In many conditions family studies have reported the rate with

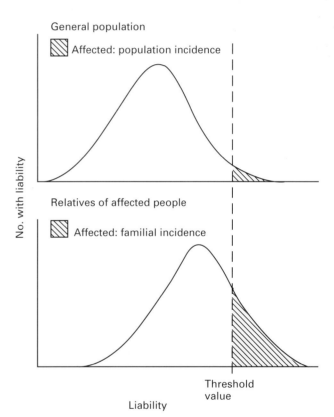

Figure 12.1 Relative contribution of environmental and genetic factors in some common disorders

Figure 12.2 Hypothetical distribution of liability of a multifactorial disorder in general population and affected families

Table 12.1 Empirical recurrence risks to siblings in Hirschsprung disease, according to sex of person affected and length of aganglionic segment

Length of colon affected	Sex of person affected	Risk to siblings (%)	
		Brothers	Sisters
Short segment	Male	4.7	0.6
	Female	8.1	2.9
Long segment	Male	16.1	11.1
	Female	18.2	9.1

which relatives of the index case have been affected. This allows empirical values for risk of recurrence to be calculated, which can be used in genetic counselling. Risks are mainly increased for first degree relatives. Second degree relatives have a slight increase in risk only and third degree relatives usually have the same risk as the general population. The severity of the disorder and the number of affected individuals in the family also affect recurrence risk. The recurrence risk for bilateral cleft lip and palate is higher than the recurrence risk for cleft lip alone, and the recurrence risk for neural tube defect is 4% after one affected child, but 12% after two. Some conditions are more common in one sex than the other. In these disorders the risk of recurrence is higher if the disorder has affected the less frequently affected sex. As with the other examples, the greater genetic susceptibility in the index case confers a higher risk to relatives. A rational approach to preventing multifactorial disease is to modify known environmental triggers in genetically susceptible subjects. Folic acid supplementation in pregnancies at increased risk of neural tube defects and modifying diet and smoking habits in coronary heart disease are examples of effective intervention, but this approach is not currently possible for many disorders.

Heritability

The heritability of a variable trait or disorder reflects the proportion of the variation that is due to genetic factors. The level of this genetic contribution to the aetiology of a disorder can be calculated from the disease incidence in the general population and that in relatives of an affected person. Disorders with a greater genetic contribution have higher heritability, and hence, higher risks of recurrence.

HLA association and linkage

Several important disorders occur more commonly than expected in subjects with particular HLA phenotypes, which implies that certain HLA determinants may affect disease susceptibility. Awareness of such associations may be helpful in counselling. For example, ankylosing spondylitis, which has an overall risk of recurrence of 4% in siblings, shows a strong association with HLA-B27, and 95% of affected people are positive for this antigen. The risk to their first degree relatives is increased to 9% for those who are also positive for HLA-B27 but reduced to less than 1% for those who are negative.

Genetic association, which may imply a causal relation, is different from genetic linkage, which occurs when two gene loci are physically close together on the chromosome. A disease gene, located near the HLA complex of genes on chromosome 6, will be linked to a particular HLA haplotype within a given affected family but will not necessarily be associated with the same HLA antigens in unrelated affected people. HLA typing can be used to predict disease by establishing the linked HLA haplotype within a given family.

Congenital adrenal hyperplasia due to 21-hydroxylase deficiency shows both linkage and association with histocompatibility antigens. The 21-hydroxylase gene lies within the HLA gene cluster and is therefore linked to the HLA haplotype. In addition, the salt-losing form of 21-hydroxylase deficiency is associated with HLA-Bw47 antigen. This combination of linkage and association is known as linkage disequilibrium and results in certain alleles at neighbouring loci occurring together more often than would be expected by chance.

Table 12.2 Estimates of heritability

	Heritability (%)
Schizophrenia	85
Asthma	80
Cleft lip and palate	76
Coronary heart disease	65
Hypertension	62
Neural tube defect	60
Peptic ulcer	37

Table 12.3 Diseases associated with histocompatibility antigens

Ankylosing spondylitis	B27
Autoimmune thyroid disease	B8, DR3
Chronic active hepatitis	B8, DR3
Coeliac disease	B8, DR3
Diabetes (juvenile)	B8, DR3
	B15, DR4
Haemochromatosis	A3
Multiple sclerosis	DR2
Psoriasis	CW6
Reiter syndrome	B27
Rheumatoid arthritis	DR4

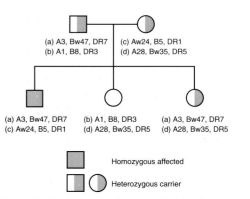

(a) A3, Bw47, DR7
(b) A1, B8, DR3

(c) Aw24, B5, DR1
(d) A28, Bw35, DR5

(a) A3, Bw47, DR7
(c) Aw24, B5, DR1

(b) A1, B8, DR3
(d) A28, Bw35, DR5

(a) A3, Bw47, DR7
(d) A28, Bw35, DR5

Homozygous affected

Heterozygous carrier

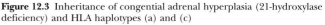

Figure 12.3 Inheritance of congenital adrenal hyperplasia (21-hydroxylase deficiency) and HLA haplotypes (a) and (c)

Twins

Twins share a common intrauterine environment, but though monozygous twins are genetically identical with respect to their inherited nuclear DNA, dizygous twins are no more alike than any other pair of siblings, sharing, on average, half their genes. This provides the basis for studying twins to determine the genetic contribution in various disorders, by comparing the rates of concordance or discordance for a particular trait between pairs of monozygous and dizygous twins. The rate of concordance in monozygous twins is high for disorders in which genetic predisposition plays a major part in the aetiology of the disease. The phenotypic variability of genetic traits can be studied in monozygous twins, and the effect of a shared intrauterine environment may be studied in dizygous twins.

Twins may be derived from a single egg (monozygous, identical) or two separate eggs (dizygous, fraternal). Examination of the placenta and membranes may help to distinguish between monozygous and dizygous twins but is not completely reliable. Monozygosity, resulting in twins of the same sex who look alike, can be confirmed by investigating inherited characteristics such as blood group markers or DNA polymorphisms (fingerprinting).

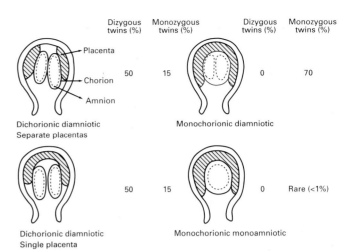

Figure 12.4 Placentation in monozygotic and dizygotic twins

Diabetes

A genetic predisposition is well recognised in both type I insulin dependent diabetes (IDDM) and type II non-insulin dependent diabetes (NIDDM). Maturity onset diabetes of the young (MODY) is a specific form of non-insulin dependent diabetes that follows autosomal dominant inheritance and has been shown to be due to mutations in a number of different genes. Clinical diabetes or impaired glucose tolerance also occurs in several genetic syndromes, for example, haemochromatosis, Friedreich ataxia, and Wolfram syndrome (diabetes mellitus, optic atrophy, diabetes insipidus and deafness). Only rarely is diabetes caused by the secretion of an abnormal insulin molecule.

IDDM affects about 3 per 1000 of the population in the UK and is a T cell dependent autoimmune disease. Genetic predisposition is important, but only 30% of monozygous twins are concordant for the disease and this indicates that environmental factors (such as triggering viral infections) are also involved. About 60% of the genetic susceptibility to IDDM is likely to be due to genes in the HLA region. The overall risk to siblings is about 6%. This figure rises to 16% for HLA identical siblings and falls to 1% if they have no shared haplotype. An association with DR3 and DR4 class II antigens is well documented, with 95% of insulin dependent diabetics having one or both antigens, compared to 50–60% of the normal population. As most people with DR3 or DR4 class II antigens do not develop diabetes, these antigens are unlikely to be the primary susceptibility determinants. Better definition of susceptible genotypes is becoming possible as subgroups of DR3 and DR4 serotypes are defined by molecular analysis. For example, low risk HLA haplotypes that confer protection always have aspartic acid at position 57 of the DQB1 allele. High risk haplotypes have a different amino acid at this position and homozygosity for non-aspartic acid residues is found much more often in diabetics than in non-diabetics.

The second locus identified for IDDM was found to be close to the insulin gene on chromosome 11. Susceptibility is dependent on the length of a 14bp minisatellite repeat unit. Short repeats (26–63 repeat units) confer susceptibility,

Box 12.2 Twinning

Dizygous twins
- May be familial
- More common in black people than white Europeans

Monozygous twins
- Seldom familial
- Occur in 0.4% of all pregnancies
- Associated with twice the risk of congenital malformations as singleton or dizygous twin pregnancies

Table 12.4 General distinction between insulin dependent and non-insulin dependent diabetes

	Insulin dependent diabetes	Non-insulin dependent diabetes
Clinical features	Thinness	Obesity
	Ketosis	No ketosis
	Early onset	Late onset
Treatment	Insulin	Diet or drugs
Concordance in monozygotic twins	30%	40–100%
Histocompatibility antigens	Associated	Not associated
Autoimmune disease	Associated	Not associated
Antibodies to insulin and islet cells	Present	Absent

Table 12.5 Empirical risk for diabetes according to affected members of family

	Risk (%)
Insulin dependent diabetes	
Sibling	1–16
One parent	4
Both parents	20
Monozygous twin	30
Non-insulin dependent diabetes	
First degree relative	10–40
Monozygous twin	40–100
Maturity onset diabetes of the young	
First degree relative	50

perhaps by influencing the expression of the insulin gene in the developing thymus. Subsequent mapping studies have identified a number of other possible IDDM susceptability loci throughout the genome, whose modes of action are not yet known.

NIDDM is due to relative insulin deficiency and insulin resistance. There is a strong genetic predisposition although other factors such as obesity are important. Concordance in monozygotic twins is 40–100% and the risk to siblings may approach 40% by the age of 80. Although the biochemical mechanisms underlying NIDDM are becoming better understood, the genetic causes remains obscure. In rare cases, insulin receptor gene mutations, mitochondrial DNA mutations or mild mutations in some of the MODY genes are thought to confer susceptibility to NIDDM.

Coronary heart disease

Environmental factors play a very important role in the aetiology of coronary heart disease, and many risk factors have been identified, including high dietary fat intake, impaired glucose tolerance, raised blood pressure, obesity, smoking, lack of exercise and stress. A positive family history is also important. The risk to first degree relatives is increased to six times above that of the general population, indicating a considerable underlying genetic predisposition. Lipids play a key role and coronary heart disease is associated with high LDL cholesterol, high ApoB (the major protein fraction of LDL), low HDL cholesterol and elevated Lp(a) lipoprotein levels. High circulating Lp(a) lipoprotein concentration has been suggested to have a population attributable risk of 28% for myocardial infarction in men aged under 60. Other risk factors may include low activity of paraoxonase and increased levels of homocysteine and plasma fibrinogen.

Lipoprotein abnormalities that increase the risk of heart disease may be secondary to dietary factors, but often follow multifactorial inheritance. About 60% of the variability of plasma cholesterol is genetic in origin, influenced by allelic variation in many genes including those for ApoE, ApoB, ApoA1 and hepatic lipase that individually have a small effect. Familial hypercholesterolaemia (type II hyperlipoproteinaemia), on the other hand, is dominantly inherited and may account for 10–20% of all early coronary heart disease. One in 500 of the general population is estimated to be heterozygous for the mutant *LDLR* gene. The risk of coronary heart disease increases with age in heterozygous subjects, who may also have xanthomas. Severe disease, often presenting in childhood, is seen in homozygous subjects.

Familial aggregations of early coronary heart disease also occur in people without any detectable abnormality in lipid metabolism. Risks to other relatives will be high, and known environmental triggers should be avoided. Future molecular genetic studies may lead to more precise identification of subjects at high risk as potential candidate genes are identified.

Schizophrenia and affective psychoses

A strong familial tendency is found in both schizophrenia and affective disorders. The importance of genetic rather than environmental factors has been shown by reports of a high incidence of schizophrenia in children of affected parents and

> **Box 12.3 Factors indicating increased risk of insulin dependent diabetes**
>
> - Insulin autoantibodies
> - Islet cell antibodies
> - Activated T lymphocytes
> - Specific HLA haplotypes

> **Box 12.4**
>
> The prevalence of non-insulin dependent diabetes (NIDDM) is increasing worldwide and it has been estimated that some 250 million people will be affected by the year 2020.

Table 12.6 Risk factors in coronary heart disease

Environmental	Genetic
• Smoking	• Family history
• Obesity	• Lipid abnormality:
• High blood pressure	LDLR
• Diet	ApoA, B, E
• Lack of exercise	Lp(a)
• Stress	

Table 12.7 Types of hyperlipidaemia

	WHO type	Excess
Autosomal dominant		
Familial hypercholesterolaemia	IIa, IIb	LDL
Familial combined hyperlipidaemia	IIa, IIb, IV	LDL, VLDL
Familial hypertriglyceridaemia	V, VI	VLDL, CM
Autosomal recessive		
Apolipoprotein CII deficiency	I, V	CM, VLDL
Polygenic		
Common hypercholesterolaemia	IIa	LDL

LDL = low density lipoprotein; VLDL = very low density lipoprotein; CM = chylomicrons

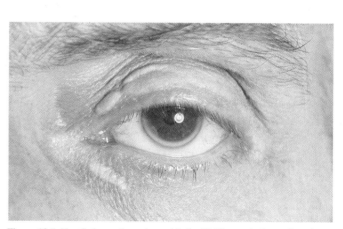

Figure 12.5 Xanthelasma in patient with familial hypercholesterolaemia

concordance in monozygotic twins, even when they are adopted and reared apart from their natural relatives. The same is true of manic depression. Empirical values for lifetime risk of recurrence are available for counselling, and the burden of the disorders needs to be taken into account. Both polygenic and single major gene models have been proposed to explain genetic susceptibility. A search for linked biochemical or molecular markers in large families with many affected members has so far failed to identify any major susceptibility genes.

Congenital malformations

Syndromes of multiple congenital abnormalities often have mendelian, chromosomal or teratogenic causes, many of which can be identified by modern cytogenetic and DNA techniques. Some malformations are non-genetic, such as the amputations caused by amniotic bands after early rupture of the amnion. Most isolated congenital malformations, however, follow multifactorial inheritance and the risk of recurrence depends on the specific malformation, its severity and the number of affected people in the family. Decisions to have further children will be influenced by the fact that the risk of recurrence is generally low and that surgery for many isolated congenital malformations is successful. Prenatal ultrasonography may identify abnormalities requiring emergency neonatal surgery or severe malformations that have a poor prognosis, but it usually gives reassurance about the normality of a subsequent pregnancy.

Mental retardation or learning disability

Intelligence is a polygenic trait. Mild learning disability (intelligence quotient 50–70) represents the lower end of the normal distribution of intelligence and has a prevalence of about 3%. The intelligence quotient of offspring is likely to lie around the mid-parental mean. One or both parents of a child with mild learning disability often have similar disability themselves and may have other learning-disabled children. Intelligent parents who have one child with mild learning disability are less likely to have another similarly affected child.

By contrast, the parents of a child with moderate or severe learning disability (intelligence quotient < 50) are usually of normal intelligence. A specific cause is more likely when the retardation is severe and may include chromosomal abnormalities and genetic disorders. The risk of recurrence depends on the diagnosis but in severe non-specific retardation is about 3% for siblings. A higher recurrence risk is observed after the birth of an affected male because some of these cases represent X linked disorders. Recurrence risks are also higher (about 15%) if the parents are consanguineous, because of the increased likelihood of an autosomal recessive aetiology. The recurrence risk for any couple increases to 25% after the birth of two affected children.

Table 12.8 Overall incidence and empirical risk of recurrence (%) in schizophrenia and affective psychosis according to affected relative

	Schizophrenia	Affective psychosis
Incidence in general population	1	2–3
Sibling	9	13
One parent	13	15
Both parents	45	50
Monozygous twin	40	70
Dizygous twin	10	20
Second degree relative	3	5
Third degree relative	1–2	2–3

Table 12.9 Risk of recurrence in siblings for some common congenital malformations

	Risk
Anencephaly or spina bifida	5*
Congenital heart disease	1–4
Cleft lip and palate	4
Cleft palate alone	2
Renal agenesis	3
Pyloric stenosis	2–10[†]
Congenital dislocated hip	1–11[†]
Club foot	3
Hypospadias	10
Cryptorchidism	10
Tracheo-oesophageal fistula	1
Exomphalos	<1

* Risk reduced by periconceptional supplementation with folic acid
[†] Risk affected by sex of index case or sibling, or both

Table 12.10 Risk of recurrence for severe non-specific mental retardation according to affected relative

	Risk
One sibling	1 in 35
male sibling	1 in 25
female sibling	1 in 50
One sibling (consanguineous parents)	1 in 7
Two siblings	1 in 4
Male sibling plus maternal uncle or male cousin	X linked

13 Dysmorphology and teratogenesis

Dysmorphology is the study of malformations arising from abnormal embryogenesis. A significant birth defect affects 2–4% of all liveborn infants and 15–20% of stillbirths. Recognition of patterns of multiple congenital malformations may allow inferences to be made about the timing, mechanism, and aetiology of structural developmental defects. Animal research is providing information about cellular interactions, migration and differentiation processes, and gives insight into the possible mechanisms underlying human malformations. Molecular studies are now identifying defects such as submicroscopic chromosomal deletions and mutations in developmental genes as the underlying cause of some recognised syndromes. Diagnosing multiple congenital abnormality syndromes in children can be difficult but it is important to give correct advice about management, prognosis and risk of recurrence.

Figure 13.1 Dysmorphic facial features and severe developmental delay in child with deletion of chromosome 1 (1p36). This chromosomal abnormality may not be detected by routine cytogenetic analysis. Recognition of clinical features and fluorescence in situ hybridisation analysis enables diagnosis

Definition of terms

Malformation

A malformation is a primary structural defect occurring during the development of an organ or tissue. Most malformations have occurred by 8 weeks of gestation. An isolated malformation, such as cleft lip and palate, congenital heart disease or pyloric stenosis, can occur in an otherwise normal child. Most single malformations are inherited as polygenic traits with a fairly low risk of recurrence, and corrective surgery is often successful. Multiple malformation syndromes comprise defects in two or more systems and many are associated with mental retardation. The risk of recurrence is determined by the aetiology, which may be chromosomal, teratogenic, due to a single gene, or unknown. Minor anomalies are those that cause no significant physical or functional effect and can be regarded as normal variants if they affect more than 4% of the population. The presence of two or more minor anomalies indicates an increased likelihood of a major anomaly being present.

Figure 13.2 Malformation: exomphalos with herniation of abdominal organs through the abdominal wall defect. Exomphalos may occur as an isolated anomaly or as part of a multiple malformation syndrome or chromosomal disorder

Disruption

A disruption defect implies that there is destruction of a part of a fetus that had initially developed normally. Disruptions usually affect several different tissues within a defined anatomical region. Amniotic band disruption after early rupture of the amnion is a well-recognised entity, causing constriction bands that can lead to amputations of digits and limbs. Sometimes more extensive disruptions occur, such as facial clefts and central nervous system defects. Interruption of the blood supply to a developing part from other causes will also cause disruption due to infarction with consequent atresia. The prognosis is determined by the severity of the physical defect. As the fetus is genetically normal and the defects are caused by an extrinsic abnormality the risk of recurrence is small.

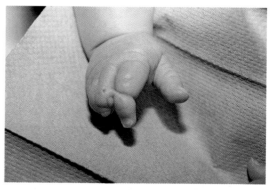

Figure 13.3 Disruption: amputation of the digits, syndactyly and constriction bands as a consequence of amniotic band disruption

Deformation

Deformations are due to abnormal intrauterine moulding and give rise to deformity of structurally normal parts. Deformations usually involve the musculoskeletal system and may occur in fetuses with underlying congenital neuromuscular problems such as spinal muscular atrophy and congenital myotonic dystrophy. Paralysis in spina bifida also gives rise to positional deformities of the legs and feet. In these disorders

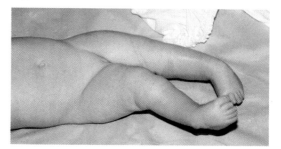

Figure 13.4 Deformation: Lower limb deformity in an infant with arthrogryposis due to amyoplasia

the prognosis is often poor and the risk of recurrence for the underlying disorder may be high.

Oligohydramnios causes fetal deformation and is well recognised in fetal renal agenesis (Potter sequence). The absence of urine production by the fetus results in severe oligohydramnios, which in turn causes fetal deformation and pulmonary hypoplasia. Oligohydramnios caused by chronic leakage of liquor has a similar effect.

A normal fetus may be constrained by uterine abnormalities, breech presentation or multiple pregnancy. The prognosis is generally excellent, and the risk of recurrence is low except in cases of structural uterine abnormality.

Dysplasia
Dysplasia refers to abnormal cellular organisation or function within a specific organ or tissue type. Most dysplasias are caused by single gene defects, and include conditions such as skeletal dysplasias and storage disorders from inborn errors of metabolism. Unlike the other mechanisms causing birth defects, dysplasias may have a progressive effect and can lead to continued deterioration of function.

Classification of birth defects

Single system defects
Single system defects constitute the largest group of birth defects, affecting a single organ system or local region of the body. The commonest of these include cleft lip and palate, club foot, pyloric stenosis, congenital dislocation of the hip and congenital heart defects. Each of these defects can also occur frequently as a component of a more generalised multiple abnormality disorder. Congenital heart defects, for example, are associated with many chromosomal disorders and malformation syndromes. When these defects occur as isolated abnormalities, the recurrence risk is usually low.

Multiple malformation syndromes
When a combination of congenital abnormalities occurs together repeatedly in a consistent pattern due to a single underlying cause, the term "syndrome" is used. The literal translation of this Greek term is "running together". Identification of a birth defect syndrome allows comparison of cases to define the clinical spectrum of the disorder and aids research into aetiology and pathogenesis.

Sequences
The term sequence implies that a series of events occurs after a single initiating abnormality, which may be a malformation, a deformation or a disruption. The features of Potter sequence are classed as a malformation sequence because the initial abnormality is renal agenesis, which gives rise to oligohydramnios and secondary deformation and pulmonary hypoplasia. Other examples are the holoprosencephaly sequence and the sirenomelia sequence. In holoprosencephaly the primary developmental defect is in the forebrain, leading to microcephaly, absent olfactory and optic nerves, and midline defects in facial development, including hypotelorism or cyclopia, midline cleft lip and abnormal development of the nose. In sirenomelia the primary defect affects the caudal axis of the fetus, from which the lower limbs, bladder, genitalia, kidneys, hindgut and sacrum develop. Abnormalities of all these structures occur in the sirenomelia sequence.

Associations
Certain malformations occur together more often than expected by chance alone and are referred to as associations.

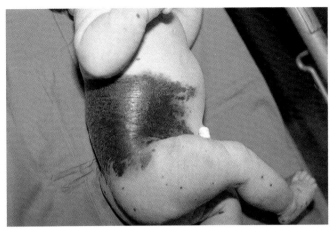

Figure 13.5 Dysplasia: giant melanocytic naevus accompanied by smaller congenital naevi usually represents a sporadic dysplasia with low recurrence risk. (courtesy of Professor Dian Donnai, Regional Genetic Service, St Mary's Hospital, Manchester)

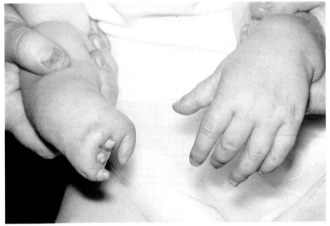

Figure 13.6 Unilateral terminal transverse defect of the hand occuring as an isolated malformation in an otherwise healthy baby

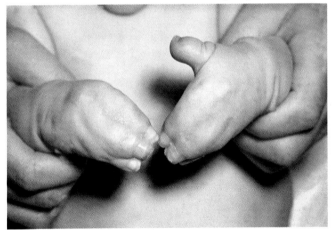

Figure 13.7 Bilateral syndactyly affecting all fingers on both hands occuring as part of Apert syndrome in a child with craniosynostosis due to a new mutation in the fibroblast growth factor receptor-2 gene

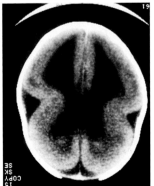

Figure 13.8 Isolated lissencephaly sequence due to neuronal migration defect is heterogeneous. Some cases are due to submicroscopic deletions of chromosome 17p involving the *LIS1* gene, others are secondary to intrauterine CMV infection or early placental insufficiency

There is great variation in clinical presentation, with different children having different combinations of the related abnormalities. The names given to recognised malformation associations are often acronyms of the component abnormalities. Hence the *Vater* association consists of *v*ertebral anomalies, *a*nal atresia, *t*racheo-*o*esophageal fistula and *r*adial defects. The acronym *vacterl* has been suggested to encompass the additional *c*ardiac, *r*enal and *l*imb defects of this association.

Murcs association is the name given to the non-random occurrence of *Mu*llerian duct aplasia, *r*enal aplasia and *c*ervicothoracic *s*omite dysplasia. In the *Charge* association the related abnormalities include *c*olobomas of the eye, *h*eart defects, choanal *a*tresia, mental *r*etardation, *g*rowth retardation and *e*ar anomalies.

Complexes

The term developmental field complex has been used to describe abnormalities that occur in adjacent or related structures from defects that affect a particular geographical part of the developing embryo. The underlying aetiology may represent a vascular event, resulting in the defects such as those seen in hemifacial microsomia (Goldenhar syndrome), Poland anomaly and some cases of Möbius syndrome.

Identification of syndromes

Recognition of multiple malformation syndromes is important to answer the questions that parents of all babies with congenital malformations ask, namely:

What is it?
Why did it happen?
What does it mean for the child's future?
Will it happen again?

Parents often experience feelings of guilt after the birth of an abnormal child, and time spent discussing what is known about the aetiology of the abnormalities may help to alleviate some of their fears. They also need an explanation of what to expect in terms of treatment, anticipated complications and long term outlook. Accurate assessment of the risk of recurrence cannot be made without a diagnosis, and the availability of prenatal diagnosis in subsequent pregnancies will depend on whether there is an associated chromosomal abnormality, a structural defect amenable to detection by ultrasonography, or an identifiable biochemical or molecular abnormality.

The assessment of infants and children with malformations requires documentation of a detailed history and a physical examination. Parental age and family history may provide clues about the aetiology. Any abnormalities during the pregnancy, including possible exposure to teratogens, should be recorded, as well as the mode of delivery and the occurrence of any perinatal problems. The subsequent general health, growth, developmental progress and behaviour of the child must also be assessed. Examination of the child should include a search for both major and minor anomalies with documentation of the abnormalities present and accurate clinical measurements and photographic records whenever possible. Investigations required may include chromosomal analysis and molecular, biochemical or radiological studies.

A chromosomal or mendelian aetiology has been identified for many multiple congenital malformation syndromes enabling appropriate recurrence risks to be given. When the aetiology of a recognised multiple malformation syndrome is not known, empirical figures for the risk of recurrence derived from family studies can be used, and these are usually fairly low. The genetic abnormality underlying de Lange syndrome,

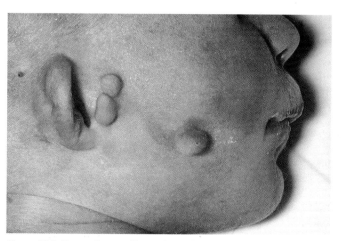

Figure 13.9 External ear malformation with preauricular skin tags in Goldenhar syndrome

Figure 13.10 The diagnosis of de Lange syndrome is based on characteristic facial features associated with growth failure and developmental delay. Some cases have upper limb anomalies

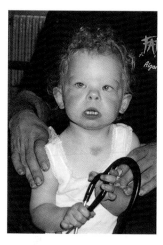

Figure 13.11 William syndrome, associated with characteristic facial appearance, developmental delay, cardiac abnormalities and infantile hypercalcaemia is due to a submicroscopic deletion of chromosome 7q, diagnosed by fluorescence in situ hybridisation analysis

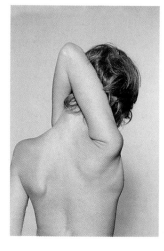

Figure 13.12 Extreme joint laxity in autosomal dominant Ehlers Danlos syndrome type 1. Some cases are due to mutations in the collagen genes *COL5A1*, *COL5A2* and *COL1A1*

for example, is not yet known, but recurrence risk is very low. Consanguineous marriages may give rise to autosomal recessive syndromes unique to a particular family. In this situation, the recurrence risk for an undiagnosed multiple malformation syndrome is likely to be high. In any family with more than one child affected, it is appropriate to explain the 1 in 4 risk of recurrence associated with autosomal recessive inheritance, although some cases may be due to a cryptic familial chromosomal rearrangement.

The molecular basis of an increasing number of birth defect syndromes is being defined, as genes involved in various processes instrumental in programming early embryonic development are identified. Mutations in the family of fibroblast growth factor receptor genes have been found in some skeletal dysplasias (achondroplasia, hypochondroplasia and thanatophoric dysplasia), as well as in a number of craniosynostosis syndromes. Other examples include mutations in the *HOXD13* gene in synpolydactyly, in the *PAX3* gene in Waardenberg syndrome type I, in the *PAX6* gene in aniridia type II, and in the *SOX9* gene in campomelic dysplasia.

Numerous malformation syndromes have been identified, and many are extremely rare. Published case reports and specialised texts often have to be reviewed before a diagnosis can be reached. Computer programs are available to assist in differential diagnosis, but despite this, malformation syndromes in a considerable proportion of children remain undiagnosed.

Stillbirths

Detailed examination and investigation of malformed fetuses and stillbirths is essential if parents are to be accurately counselled about the cause of the problem, the risk of recurrence, and the availability of prenatal tests in future pregnancies. As with liveborn infants, careful documentation of the abnormalities is required with detailed photographic records. Cardiac blood samples and skin or cord biopsy specimens should be taken for chromosomal analysis and bacteriological and virological investigations performed. Other investigations, including full skeletal *x* ray examination and tissue sampling for biochemical studies and DNA extraction, may be necessary. Autopsy will determine the presence of associated internal abnormalities, which may permit diagnosis.

Environmental teratogens

Drugs

Identification of drugs that cause fetal malformations is important as they constitute a potentially preventable cause of abnormality. Although fairly few drugs are proved teratogens in humans, and some drugs are known to be safe, the accepted policy is to avoid all drugs if possible during pregnancy. Thalidomide has been the most dramatic teratogen identified, and an estimated 10 000 babies worldwide were damaged by this drug in the early 1960s before its withdrawal.

Alcohol is currently the most common teratogen, and studies suggest that between 1 in 300 and 1 in a 1000 infants are affected. In the newborn period, exposed infants may have tremulousness due to withdrawal, and birth defects such as microcephaly, congenital heart defects and cleft palate. There is often a characteristic facial appearance with short palpebral fissures, a smooth philtrum and a thin upper lip. Children with the fetal alcohol syndrome exhibit prenatal and postnatal growth deficiency, developmental delay with subsequent learning disability, and behavioural problems.

Treatment of epilepsy during pregnancy presents a particular problem, as 1% of pregnant women have a

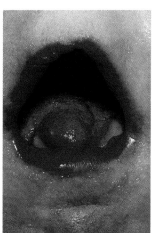

Figure 13.13 Lobulated tongue in orofaciodigital syndrome type 1 (OFD 1) inherited in an X-linked dominant fashion due to mutations in the *CXORF5* gene

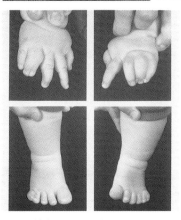

Figure 13.14 Hand and foot abnormalities in synpolydactyly due to autosomal dominant mutation in the *HOXD13* gene (courtesy of Professor Dian Donnai, Regional Genetic Service, St. Mary's Hospital Manchester)

Figure 13.15 Thanatophoric dysplasia: usually sporadic lethal bone dysplasia due to mutations in the fibroblast growth factor receptor-3 gene (courtesy of Professor Dian Donnai, Regional Genetic Service, St. Mary's Hospital, Manchester)

Figure 13.16 Limb malformation due to intrauterine exposure to thalidomide (courtesy of Professor Dian Donnai, Regional Genetic Service, St Mary's Hospital, Manchester)

seizure disorder and all anticonvulsants are potentially teratogenic. There is a two to three-fold increase in the incidence of congenital abnormalities in infants of mothers treated with anticonvulsants during pregnancy. Recognisable syndromes, often associated with learning disability, occur in a proportion of pregnancies exposed to phenytoin and sodium valproate. An increased risk of neural tube defect has been documented with sodium valproate and carbamazepine therapy, and periconceptional supplementation with folic acid is advised. Anticonvulsant therapy during pregnancy may be essential to prevent the risks of grand mal seizures or status epilepticus. Whenever possible monotherapy using the lowest effective therapeutic dose should be employed.

Maternal disorders

Several maternal disorders have been identified in which the risk of fetal malformations is increased, including diabetes and phenylketonuria. The risk of congenital malformations in the pregnancies of diabetic women is two to three times higher than that in the general population but may be lowered by good diabetic control before conception and during the early part of pregnancy. In phenylketonuria the children of an affected woman will be healthy heterozygotes in relation to the abnormal gene, but if the mother is not returned to a carefully controlled diet before pregnancy the high maternal serum concentration of phenylalanine causes microcephaly in the developing fetus.

Intrauterine infection

Various intrauterine infections are known to cause congenital malformations in the fetus. Maternal infection early in gestation may cause structural abnormalities of the central nervous system, resulting in neurological abnormalities, visual impairment and deafness, in addition to other malformations, such as congenital heart disease. When maternal infection occurs in late pregnancy the risk that the infective agent will cross the placenta is higher, and the newborn infant may present with signs of active infection, including hepatitis, thrombocytopenia, haemolytic anaemia and pneumonitis.

Rubella embryopathy is well recognised, and the aim of vaccination programmes against rubella-virus during childhood is to reduce the number of non-immune girls reaching childbearing age. The presence of rubella-specific IgM in fetal or neonatal blood samples identifies babies infected in utero. Cytomegalovirus is a common infection and 5–6% of pregnant women may become infected. Only 3% of newborn infants, however, have evidence of cytomegalovirus infection, and no more than 5% of these develop subsequent problems. Infection with cytomegalovirus does not always confer natural immunity, and occasionally more than one sibling has been affected by intrauterine infection. Unlike for rubella, vaccines against cytomegalovirus or toxoplasma are not available, and although active maternal toxoplasmosis can be treated with drugs such as pyrimethamine, this carries the risk of teratogenesis.

Herpes simplex infection in the newborn infant is generally acquired at the time of birth, but infection early in pregnancy is probably associated with an increased risk of abortion, late fetal death, prematurity and structural abnormalities of the central nervous system. Maternal varicella infection may also affect the fetus, causing abnormalities of the central nervous system and cutaneous scars. The risk of a fetus being affected by varicella infection is not known but is probably less than 10%, with a critical period during the third and fourth months of pregnancy. Affected infants seem to have a high perinatal mortality rate.

Figure 13.17 Children exposed to sodium valproate in utero may develop fetal anticonvulsant syndrome associated with facial dysmorphism (note thin upper lip and smooth philtrum), congenital malformations (spina bifida, cleft lip and palate and congenital heart defects), learning disability and behavioural problems

Box 13.1 Examples of teratogens

Drugs
- Alcohol
- Anticonvulsants
  phenytoin
  sodium valproate
  carbamazepine
- Anticoagulants
  warfarin
- Antibiotics
  streptomycin
- Treatment for acne
  tetracycline
  isotretinoin
- Antimalarials
  pyrimethamine
- Anticancer drugs
- Androgens

Environmental chemicals
- Organic mercurials
- Organic solvents

Ionizing radiation

Maternal disorders
- Epilepsy
- Diabetes
- Phenylketonuria
- Hyperpyrexia
- Iodine deficiency

Intrauterine infections
- Rubella
- Cytomegalovirus
- Toxoplasmosis
- Herpes simplex
- Varicella zoster
- Syphilis

14   Prenatal diagnosis

Prenatal diagnosis is important in detecting and preventing genetic disease. Significant advances since the mid-1980s have been the development of chorionic villus sampling procedures in the first trimester and the application of recombinant DNA techniques to the diagnosis of many mendelian disorders. Techniques for undertaking diagnosis on single cells has more recently made preimplantation diagnosis of some genetic disorders possible. Various prenatal procedures are available, generally being performed between 10 and 20 weeks' gestation. Having prenatal tests and waiting for results is stressful for couples. They must be supported during this time and given full explanation of results as soon as possible. Most tertiary centres have developed fetal management teams consisting of obstetricians, midwives, radiologists, neonatologists, paediatric surgeons, clinical geneticists and counsellors, to provide integrated services for couples in whom prenatal tests detect an abnormality.

Indications for prenatal diagnosis

Prenatal diagnosis occasionally allows prenatal treatment to be instituted but is generally performed to permit termination of pregnancy when a fetal abnormality is detected, or to reassure parents when a fetus is unaffected. Since an abnormal result on prenatal testing may lead to termination this course of action must be acceptable to the couple. Careful assessment of their attitudes is important, and all couples who elect for termination following an abnormal test result need counselling and psychological support afterwards. Couples who would not contemplate termination may still request a prenatal diagnosis to help them to prepare for the outcome of the pregnancy, and these requests should not be dismissed. The risk of the disorder occurring and its severity influence a couple's decision to embark on testing, as does the accuracy, timing and safety of the procedure itself.

Identifying risk

Pregnancies at risk of fetal abnormality may be identified in various ways. A pregnancy may be at increased risk of Down syndrome or other chromosomal abnormality because the couple already have an affected child, because of abnormal results of biochemical screening, or because of advanced maternal age. The actual risk is usually low, but prenatal testing is often appropriate, since this allows most pregnancies to continue with less anxiety. There is a higher risk of a chromosomal abnormality in the fetus when one of the parents is known to carry a familial chromosome translocation or when congenital abnormalities have been identified by prenatal ultrasound scanning. In other families, a high risk of a single gene disorder may have been identified through the birth of an affected relative. Couples from certain ethnic groups, whose pregnancies are at high risk of particular autosomal recessive disorders, such as the haemoglobinopathies or Tay–Sachs disease, can be identified before the birth of an affected child by population screening programmes. Screening for carriers of cystic fibrosis is also possible, but not generally undertaken on a population basis. In many mendelian disorders, particularly autosomal dominant disorders of late onset and X linked recessive disorders, family studies are needed to assess the risk to the pregnancy and to determine the feasibility of prenatal

Table 14.1 Techniques for prenatal diagnosis

Ultrasonography
- Safe
- Performed mainly in second trimester

Amniocentesis
- Procedure risk 0.5–1.0%
- Performed in second trimester
- Widely available

Chorionic villus sampling
- Procedure risk 1–2%
- Performed in first trimester
- Specialised technique

Cordocentesis
- Procedure risk 1%
- Performed in second trimester
- Specialised technique

Fetal tissue biopsy
- Procedure risk <3%
- Performed in second trimester
- Very specialised technique
- Limited application

Embryo biopsy
- Limited availability and application

Box 14.1 General criteria for prenatal diagnosis
- High genetic risk
- Severe disorder
- Treatment not available
- Reliable prenatal test available
- Acceptable to parents

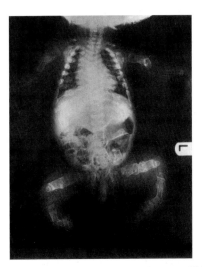

Figure 14.1 Osteogenesis imperfecta type II (perinatally lethal) can be detected by ultrasonography in the second trimester. Most cases are due to new autosomal dominant mutations but recurrence risk is around 5% because of the possibility of gonadal mosaicism in one of the parents

diagnosis before any testing procedure is performed during pregnancy.

Severity of the disorder

Several important factors must be carefully considered before prenatal testing, one of which is the severity of the disorder. For many genetic diseases this is beyond doubt; some disorders lead inevitably to stillbirth or death in infancy or childhood. Requests for prenatal diagnosis in these situations are high. The decision to terminate an affected pregnancy may be easier to make if there is no chance of the baby having prolonged survival. Equally important, however, are conditions that result in children surviving with severe, multiple, and often progressive, physical and mental handicaps, such as Down syndrome, neural tube defects, muscular dystrophy and many of the multiple congenital malformation syndromes. Again, most couples are reluctant to embark upon another pregnancy in these cases without prenatal diagnosis. Termination of pregnancy is not always the consequence of an abnormal prenatal test result. Some couples wish to know whether their baby is affected so that they can prepare themselves for the birth and care of an affected child.

Treatment for the disorder

It is also important to consider the availability of treatment for conditions amenable to prenatal diagnosis. When treatment is effective, termination may not be appropriate and invasive prenatal tests are generally not indicated, unless early diagnosis permits more rapid institution of treatment resulting in a better prognosis. Phenylketonuria, for example, can be treated effectively after diagnosis in the neonatal period, and prenatal diagnosis, although possible for parents who already have an affected child, may be inappropriate. Postnatal treatment for congenital adrenal hyperplasia due to 21-hydroxylase deficiency is also available and some couples will choose not to terminate affected pregnancies. However, in this condition, affected female fetuses become masculinised during pregnancy and have ambiguous genitalia at birth requiring reconstructive surgery. This virilisation can be prevented by starting treatment with steroids in the first trimester of pregnancy. Because of this, it may be appropriate to undertake prenatal tests to identify those pregnancies where treatment needs to continue and those where it can be safely discontinued. Prenatal diagnosis by non-invasive ultrasound scanning of major congenital malformations amenable to surgical correction is also important, as it allows the baby to be delivered in a unit with facilities for neonatal surgery and intensive care.

Test reliability

A prenatal test must be sufficiently reliable to permit decisions to be made once results are available. Some conditions can be diagnosed with certainty, others cannot, and it is important that couples understand the accuracy and limitations of any tests being undertaken. Chromosomal analysis usually provides results that are easily interpreted. Occasionally there may be difficulties, because of mosaicism or the detection of an unusual abnormality. In some cases, an abnormality other than the one being tested for will be identified, for example a sex chromosomal abnormality may be detected in a pregnancy being tested for Down syndrome. For many mendelian disorders biochemical tests or direct mutation analysis is possible. The biochemical abnormality or the presence of a mutation in an affected person or obligate carrier in the family needs to be confirmed prior to prenatal testing. Once this has been done, prenatal diagnosis or exclusion of these conditions is highly accurate. In other inherited disorders, neither

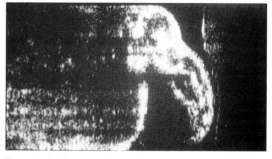

Figure 14.2 Shortened limb in Saldino–Noonan syndrome: an autosomal recessive lethal skeletal dysplasia (courtesy of Dr Sylvia Rimmer, Radiology department, St Mary's Hospital, Manchester)

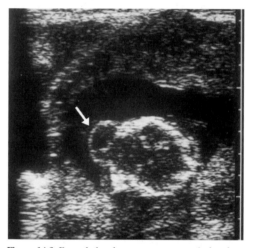

Figure 14.3 Encephalocele may represent an isolated neural tube defect or be part of a multiple malformation syndrome such as Meckel syndrome (cleft lip or palate, polydactyly, renal cystic disease and eye defects). (courtesy of Dr Sylvia Rimmer, Radiology department, St Mary's Hospital, Manchester)

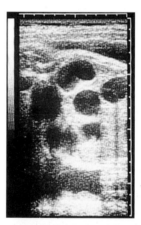

Figure 14.4 Dilated loops of bowel due to jejunal atresia, indicating the need for neonatal surgery. (courtesy of Dr Sylvia Rimmer, Radiology department, St Mary's Hospital, Manchester)

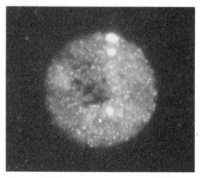

Figure 14.5 Fluorescence in situ hybridisation in interphase nuclei using chromosome 21 probes enables rapid and reliable detection of trisomy 21 (courtesy of Dr Lorraine Gaunt, Regional Genetic Service, St Mary's Hospital, Manchester)

biochemical analysis nor direct mutation testing is possible. DNA analysis using linked markers may enable a quantified risk to be given rather than an absolute result.

Screening tests

Screening tests aim to detect common abnormalities in pregnancies that are individually at low risk and provide reassurance in most cases. There is widespread application of routine screening tests for Down syndrome and neural tube defects by biochemical testing and for fetal abnormality by ultrasound scanning. Most couples will have little knowledge of the disorders being tested for and will not be anticipating an abnormal outcome at the time of testing, unlike couples undergoing specific tests for a previously recognised risk of a particular disorder. It is very important to provide information before screening so that couples know what is being tested for and appreciate the implications of an abnormal result, so that they can make an informed decision about having the tests. When abnormalities are detected, arrangements need to be made to give the results in an appropriate setting, providing sufficient information for the couple to make fully informed decisions, with continuing support from clinical staff who have experience in dealing with these situations.

Methods of prenatal diagnosis

Maternal serum screening

Estimation of maternal serum α fetoprotein (AFP) concentration in the second trimester is valuable in screening for neural tube defects. A raised AFP level indicates the need for further investigation by amniocentesis or ultrasound scanning. In some centres amniocentesis has been replaced largely by high resolution ultrasound scanning, which detects over 95% of affected fetuses.

In 1992 a combination of maternal serum AFP, β human chorionic gonadotrophin (HCG) and unconjugated estriol (uE3) in the second trimester was shown to be an effective screening test for Down syndrome, providing a composite risk figure taking maternal age into account. When 5% of women were selected for diagnostic amniocentesis following serum screening, the detection rate for Down syndrome was at least 60%, well in excess of the detection rate achieved by offering amniocentesis on the basis of maternal age alone. Serum screening does not provide a diagnostic test for Down syndrome, since the results may be normal in affected pregnancies and relatively few women with abnormal serum screening results actually have an affected fetus. Serum screening for Down syndrome is now in widespread use and diagnostic amniocentesis is generally offered if the risk of Down syndrome exceeds 1 in 250. Screening strategies include combinations of first trimester measurement of pregnancy associated plasma protein A(PAPP-A) and HCG, second trimester measurement of AFP, HCG, uE3 and inhibition A and first trimester nuchal translucency measurement.

The isolation of circulating fetal cells, such as nucleated red cells and trophoblasts from maternal blood offers a potential method for detecting genetic disorders in the fetus by a non-invasive procedure. This method could play an important role in prenatal screening for aneuploidy in the fetus, either as an independent test, or more likely, in conjunction with other tests such as ultrasonography and biochemical screening.

Ultrasonography

Obstetric indications for ultrasonography are well established and include confirmation of viable pregnancy, assessment of

Table 14.2 Applications of prenatal diagnosis

Maternal serum screening
- α Fetoprotein estimation
- Estriol and human chorionic gonadotrophin estimation

Ultrasonography
- Structural abnormalities

Amniocentesis
- α Fetoprotein and acetylcholinesterase
- Chromosomal analysis
- Biochemical analysis

Chorionic villus sampling
- DNA analysis
- Chromosomal analysis
- Biochemical analysis

Fetal blood sampling
- Chromosomal analysis
- DNA analysis

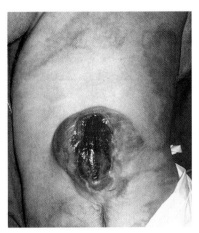

Figure 14.6 Large lumbar meningomyelocele

Box 14.2 Some causes of increased maternal serum α fetoprotein concentration

- Underestimated gestational age
- Threatened abortion
- Multiple pregnancy
- Fetal abnormality
  Anencephaly
  Open neural tube defect
  Anterior abdominal wall defect
  Turner syndrome
  Bowel atresia
  Skin defects
- Maternal hereditary persistence of α fetoprotein
- Placental haemangioma

gestational age, localisation of the placenta, assessment of amniotic fluid volume and monitoring of fetal growth. Ultrasonography is an integral part of amniocentesis, chorionic villus sampling and fetal blood sampling, and provides evaluation of fetal anatomy during the second and third trimesters.

Disorders such as neural tube defects, severe skeletal dysplasias, abdominal wall defects and renal abnormalities may all be detected by ultrasonography between 17 and 20 weeks' gestation. Centres specialising in high resolution ultrasonography can detect an increasing number of other abnormalities, such as structural abnormalities of the brain, various types of congenital heart disease, clefts of the lip and palate and microphthalmia. For some fetal malformations the improved resolution of high frequency ultrasound transducers has even enabled detection during the first trimester by transvaginal sonography. Other malformations, such as hydrocephalus, microcephaly and duodenal atresia may not manifest until the third trimester.

Abnormalities may be recognised during routine scanning of pregnancies not known to be at increased risk. In these cases it may not be possible to give a precise prognosis. The abnormality detected, for example cleft lip and palate may be an isolated defect with a good prognosis or may be associated with additional abnormalities that cannot be detected before birth in a syndrome carrying a poor prognosis. Depending on the type of abnormality detected, termination of pregnancy may be considered, or plans made for the neonatal management of disorders amenable to surgical correction.

Most single congenital abnormalities follow multifactorial inheritance and carry a low risk of recurrence, but the safety of scanning provides an ideal method of screening subsequent pregnancies and usually gives reassurance about the normality of the fetus. Syndromes of multiple congenital abnormalities may follow mendelian patterns of inheritance with high risks of recurrence. For many of these conditions, ultrasonography is the only available method of prenatal diagnosis.

Amniocentesis

Amniocentesis is a well established and widely available method for prenatal diagnosis. It is usually performed at 15 to 16 weeks' gestation but can be done a few weeks earlier in some cases. It is reliable and safe, causing an increased risk of miscarriage of around 0.5–1.0%. Amniotic fluid is aspirated directly, with or without local anaesthesia, after localisation of the placenta by ultrasonography. The fluid is normally clear and yellow and contains amniotic cells that can be cultured. Contamination of the fluid with blood usually suggests puncture of the placenta and may hamper subsequent analysis. Discoloration of the fluid may suggest impending fetal death.

The main indications for amniocentesis are for chromosomal analysis of cultured amniotic cells in pregnancies at increased risk of Down syndrome or other chromosomal abnormalities and for estimating α fetoprotein concentration and acetylcholinesterase activity in amniotic fluid in pregnancies at increased risk of neural tube defects, although few amniocenteses are now done for neural tube defects because of improved detection by ultrasonography. In specific cases biochemical analysis of amniotic fluid or cultured cells may be required for diagnosing inborn errors of metabolism. Tests on amniotic fluid usually yield results within 7–10 days, whereas those requiring cultured cells may take around 2–4 weeks. Results may not be available until 18 weeks' gestation or later, leading to late termination in affected cases.

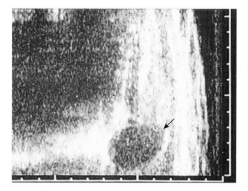

Figure 14.7 Large lumbosacral meningocele (courtesy of Dr Sylvia Rimmer, Radiology Dept, St Mary's Hospital, Manchester)

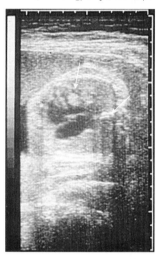

Figure 14.8 Cardiac leiomyomas in tuberous sclerosis (courtesy of Dr Sylvia Rimmer, Radiology Dept, St Mary's Hospital, Manchester)

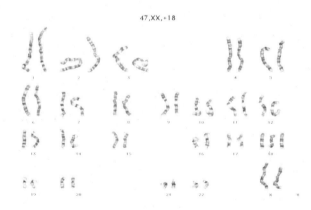

Figure 14.9 Amniocentesis procedure

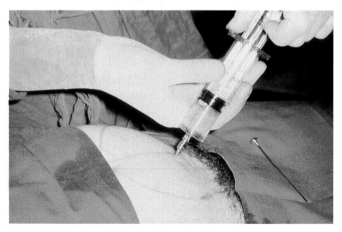

47,XX,+18

Figure 14.10 Trisomy 18 karyotype deteced by analysis of cultured amniotic cells (courtesy of Dr Lorraine Gaunt and Helena Elliott, Regional Genetic Service, St Mary's Hospital, Manchester)

Chorionic villus sampling

Chorionic villus sampling is a technique in which fetally derived chorionic villus material is obtained transcervically with a flexible catheter between 10 and 12 weeks' gestation or by transabdominal puncture and aspiration at any time up to term. Both methods are performed under ultrasound guidance, and fetal viability is checked before and after the procedure. The risk of miscarriage related to sampling in the first trimester in experienced hands is probably about 1–2% higher than the rate of spontaneous abortions at this time.

Dissection of fetal chorionic villus material from maternal decidua permits analysis of the fetal genotype. The main indications for chorionic villus sampling include the diagnosis of chromosomal disorders from familial translocations and an increasing number of single gene disorders amenable to diagnosis by biochemical or DNA analysis. The advantage of this method of testing is the earlier timing of the procedure, which allows the result to be available by about 12 weeks' gestation in many cases, with earlier termination of pregnancy, if required. These advantages have led to an increased demand for the procedure in preference to amniocentesis, particularly when the risk of the disorder occurring is high. If prenatal diagnosis is to be achieved in the first trimester it is essential to identify high risk situations and counsel couples before pregnancy so that appropriate arrangements can be made and, when necessary, supplementary family studies organised.

Fetal blood and tissue sampling

Fetal blood samples can be obtained directly from the umbilical cord under ultrasound guidance. Blood sampling enables rapid fetal karyotyping in cases presenting late in the second trimester. Indications for fetal blood sampling to diagnose genetic disorders are decreasing with the increased application of DNA analysis performed on chorionic villus material. Fetal skin biopsy has proved effective in the prenatal diagnosis of certain skin disorders and fetal liver biopsy has been performed for diagnosis of ornithine transcarbamylase (otc) deficiency. Again, the need for tissue biopsy is now largely replaced by DNA analysis on chorionic villus material and fetoscopy for direct visualisation of the fetus has been replaced by ultrasonography.

Preimplantation genetic diagnosis

Preimplantation embryo biopsy is now technically feasible for some genetic disorders and available in a few specialised centres. In this method in vitro fertilisation and embryo culture is followed by biopsy of one or two outer embryonal cells at the 6–10 cell stage of development. DNA analysis of a single cell or chromosomal analysis by in situ hybridisation is performed so that only embryos free of a particular genetic defect are reimplanted. An average IVF cycle may produce 10–15 eggs, of which five or six develop to the stage where biopsy is possible. The reported rate of pregnancy is about 20% per cycle and confirmatory genetic testing by chorionic villus biopsy or amniocentesis is recommended for established pregnancies. This method may be more acceptable to some couples than other forms of prenatal diagnosis, but has a very limited availability.

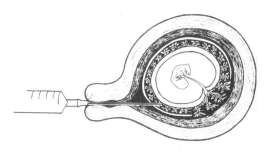

Figure 14.11 Procedure for transcervical chorionic villus sampling

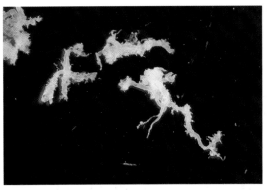

Figure 14.12 Chorionic villus material

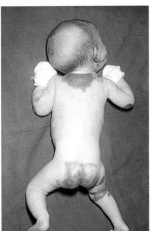

Figure 14.13 Lethal form of autosomal recessive epidermolysis bullosa, diagnosed by fetal skin biopsy if DNA analysis is not possible

Box 14.3 Potential applications of preimplantation genetic diagnosis

- Fetal sexing for X linked disorders, for example
    Duchenne muscular dystrophy
    Haemophilia
    Hunter syndrome
    Menke syndrome
    Lowe oculocerebrorenal syndrome
- Chromosomal analysis:
    Autosomal trisomies (21, 18 and 13)
    Familial chromosomal rearrangements
- Direct mutation analysis:
    Cystic fibrosis
    Childhood onset spinal muscular atrophy
    Huntington disease
    Myotonic dystrophy
    β thalassaemia
    Sickle cell disease

15 DNA structure and gene expression

The DNA molecule is fundamental to cell metabolism and cell division and it is also the basis for inherited characteristics. The central dogma of molecular genetics is the process of transferring genetic information from DNA to RNA, resulting in the production of polypeptide chains that are the basis of all proteins. Human molecular biology studies this process and its alterations in relation to health and disease. Nucleic acid, initially called nuclein, was discovered by Friedrich Miescher in 1869, but it was not until 1953 that Watson and Crick produced their model for the double helical structure of DNA and proposed the mechanism for DNA replication. During the 1960s the genetic code was found to reside in the sequence of nucleotides comprising the DNA molecule; a group of three nucleotides coding for an amino acid. The rapid expansion of molecular techniques in the past few decades has led to a better understanding of human genetic disease. The structure and function of many genes has been elucidated and the molecular pathology of various disorders is now well defined.

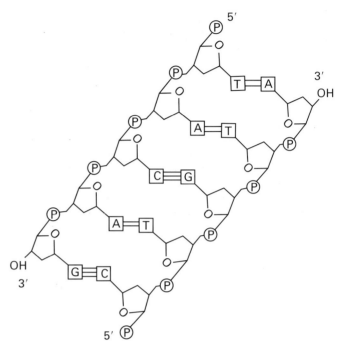

Figure 15.1 DNA molecule comprising sugar and phosphate backbone with paired nucleotides joined by hydrogen bonds

DNA and RNA structure

The linear backbone of DNA (deoxyribonucleic acid) and RNA (ribonucleic acid) consists of sugar units linked by phosphate groups. In DNA the sugar is deoxyribose and in RNA it is ribose. The orientation of the phosphate groups defines the 5′ and 3′ ends of the molecules. A nitrogenous base is attached to a sugar and phosphate group to form a nucleotide that constitutes the basic repeat unit of the DNA and RNA strands. The bases are divided into two classes: purines and pyramidines. In DNA the purines bases are adenosine (A) and guanine (G), and the pyramidine bases are cytosine (C) and thymine (T). The order of the bases along the molecule constitutes the genetic code in which the coding unit or codon, consists of three nucleotides. In RNA the arrangement of bases is the same except that thymine (T) is replaced by uracil (U).

In the nucleus, DNA exists as a double stranded helix in which the order of bases on one strand is complementary to that on the other. The bases are held together by hydrogen bonds, which allow the strands to separate and rejoin. Hydrogen bonds also contribute to the three-dimensional structure of the molecule and permit formation of RNA–DNA duplexes that are crucial for gene expression. In the DNA molecule adenine (A) is always paired with thymine (T) on the opposite strand and cytosine (C) with guanine (G). This specific pairing is fundamental to DNA replication during which the two DNA strands separate, and each acts as a template for the synthesis of a new strand, maintaining the genetic code during cell division. A similar process is used to repair and reconstitute damaged DNA. As the new DNA helix contains an existing and a newly synthesised strand the process is called semi-conservative replication. The study of cultured cells indicates that the process of cellular DNA replication takes eight hours to complete.

Transcription

Gene expression is mediated by RNA, which is synthesised using DNA as a template. This process of transcription occurs

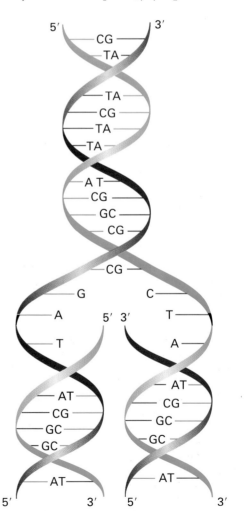

Figure 15.2 Double stranded DNA helix and semiconservative DNA replication

78

in a similar fashion to that of DNA replication. The DNA helix unwinds and one strand acts as a template for RNA transcription. RNA polymerase enzymes join ribonucleosides together to form a single stranded RNA molecule. The base sequence along the RNA molecule, which determines how the protein is made, is complementary to the template DNA strand and the same as the other, non-template, DNA strand. The non-template strand is therefore referred to as the sense strand and the template strand as the anti-sense strand. When the DNA sequence of a gene is given it relates to that of the sense strand (from 5′ to 3′ end) rather than the anti-sense strand.

The process of RNA transcription is under the control of DNA sequences in the immediate vicinity of the gene that bind transcription factors to the DNA. Once transcribed, RNA molecules undergo a number of structural modifications necessary for function, that include adding a specialised nucleoside to the 5′ end (capping) and a poly(A) tail to the 3′ end (polyadenylation). The removal of unwanted internal segments by splicing produces mature RNA. This process occurs in complexes called spliceosomes that consist of several types of snRNA (small nuclear RNA) and many proteins. Several classes of RNA are produced: mRNA (messenger RNA) directs polypeptide synthesis; tRNA (transfer RNA) and rRNA (ribosomal RNA) are involved in translation of mRNA and snRNA is involved in splicing.

In experimental systems the reverse reaction to transcription – the synthesis of complementary DNA (cDNA) using mRNA as a template – can be achieved using reverse transcriptase enzyme. This has proved to be an immensely valuable procedure for investigating human genetic disorders as it allows production of cDNA that corresponds exactly to the coding sequence of a human gene.

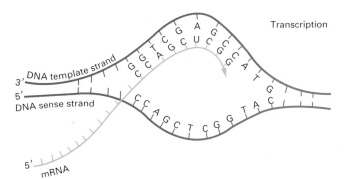

Figure 15.3 Transcription of DNA template strand

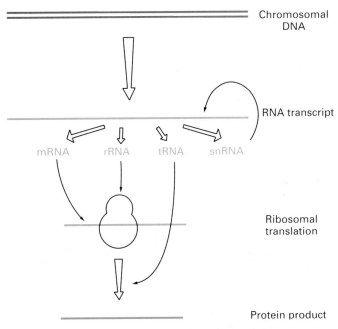

Figure 15.4 Role of different RNA molecules in the translation process

The genetic code

The basis of the genetic code lies in the order of bases along the RNA molecule. A group of three nucleotides constitute the coding unit and is referred to as the codon. Each codon specifies a particular amino acid enabling correct polypeptide assembly during protein production. The four bases in nucleic acid give 64 possible codon combinations. As there are only 20 amino acids, most are specified by more than one codon and the genetic code is therefore said to be degenerate. Some amino acids, such as methionine and tryptophan have only one codon. Others, such as leucine and serine are specified by six different codons. The third base is often involved in the degeneracy of the code, for example glycine is encoded by the triplet GGN, where N can be any base. Certain codons act to initiate or terminate polypeptide chain synthesis. The RNA triplet AUG codes for methionine and acts as a signal to start synthesis; the triplets UAA, UAG and UGA represent termination (stop) codons.

Although there are 64 codons in mRNA, there are only 30 types of cytoplasmic tRNA and 22 types of mitochondrial tRNA. To enable all 64 codons to be translated, exact nucleotide matching between the third base of the tRNA anticodon triplet and the RNA codon is not required.

The genetic code is universal to all organisms, with the exception of the mitochondrial protein production system in which four codons are differently interpreted. This alters the number of codons for four amino acids and creates an additional stop codon in the mitochondrial coding system.

Table 15.1 Genetic code (RNA)*

First base (5′ end)	Second base				Third base (3′ end)
	U	C	A	G	
U	Phe	Ser	Tyr	Cys	U
	Phe	Ser	Tyr	Cys	C
	Leu	Ser	Stop	Stop	A
	Leu	Ser	Stop	Trp	G
C	Leu	Pro	His	Arg	U
	Leu	Pro	His	Arg	C
	Leu	Pro	Gln	Arg	A
	Leu	Pro	Gln	Arg	G
A	Ile	Thr	Asn	Ser	U
	Ile	Thr	Asn	Ser	C
	Ile	Thr	Lys	Arg	A
	Met	Thr	Lys	Arg	G
G	Val	Ala	Asp	Gly	U
	Val	Ala	Asp	Gly	C
	Val	Ala	Glu	Gly	A
	Val	Ala	Glu	Gly	G

*Uracil (U) replaces thymine (T) in RNA.

Translation

After processing, mature mRNA migrates to the cytoplasm where it is translated into a polypeptide product. At either end of the mRNA molecule are untranslated regions that bind and stabilise the RNA but are not translated into the polypeptide. The translation process occurs in association with ribosomes that are composed of rRNA and protein complexes. The assembly of polypeptide chains occurs by the decoding of the mRNA triplets via tRNAs that bind specific amino acids and have an anticodon sequence that enables them to recognise an mRNA codon. Peptide bonds form between the amino acids as the tRNAs are sequentially aligned along the mRNA and translation continues until a stop codon is reached.

The process of protein production in mitochondria is similar, with mtDNA producing its own mitochondrial mRNA, tRNA and rRNA. The proteins produced in the mitochondria combine with proteins produced by nuclear genes to form the functional proteins of the mitochondrial complexes.

The primary polypeptide chains produced by the translation process undergo a variety of modifications that include chemical modification, such as phosphorylation or hydroxylation, addition of chemical groups such as carbohydrates or lipids, and internal cleavage to generate smaller mature products or to remove signal sequences in proteins once they have been secreted or transported across intracellular membranes. Many polypeptides subsequently combine with others to form the subunits of functionally active multiple protein complexes.

Gene structure and expression

The coding sequence of a gene is not continuous, but is interrupted by varying numbers and lengths of intervening non-coding sequences whose function, if any, is not known. The coding sequences are called exons and the intervening sequences introns. Human genes vary considerably in their size and complexity. A few genes, for example, the histone and glycerol kinase genes contain only one exon and no non-coding DNA, but most contain both exons and introns. Some genes contain an emormous number of exons, for example, there are 118 exons in the collagen *7A1* gene. Generally the variation between small and large genes is due to the number and size of the introns. The dystrophin gene is one of largest genes identified. It spans 2.4 million base pairs of genomic DNA, contains 79 exons and takes 16 hours to transcribe into mRNA. As with other large genes, the intronic sequences are very long and mature dystrophin mRNA is only 16kb in length (less than 1% of the genomic DNA length).

In addition to the introns, there are non-coding regions of DNA at both 5′ and 3′ ends of genes and regulatory sequences in and around the gene that control its expression. In the 5′ promoter region are two conserved, or consensus, sequences known as the TATA box and the CG or CAAT boxes. The TATA box is found in genes that are expressed only at certain times in development or in specific tissues, and the CG or CAAT boxes determine the efficiency of gene promoter activity. Other enhancer or silencer sequences at variable sites contribute to regulation of gene expression as does methylation of cytosine nucleotides, with gene expression being silenced by methylation of DNA in the promoter region.

Both coding and non-coding sequences in a gene are transcribed into mRNA. The sequences corresponding to the introns are then cut out and the exon-related sequences are spliced together to produce mature mRNA. Conserved

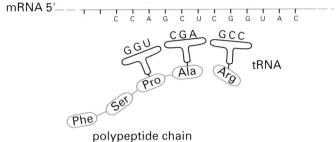

Figure 15.5 Translation of messenger RNA into protein product

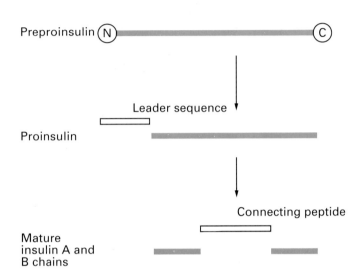

Figure 15.6 Post-translational modification of insulin

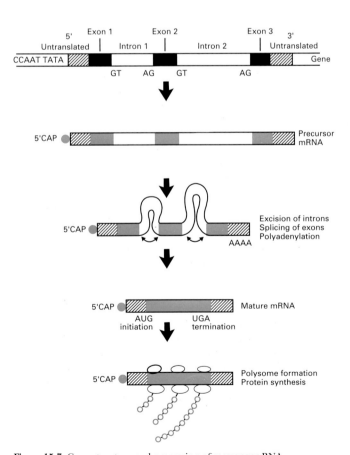

Figure 15.7 Gene structure and processing of messenger RNA

sequences at the splice sites enable their recognition in this complex process. Some genes have several different promoters that direct mRNA transcription from different initiating exons. This, together with alternative splicing, enables the production of several different isoforms of a protein from a single gene. These isoforms may be expressed in different tissues and have varying function.

Genome organisation

The term "human genome" refers to the total genetic information represented by DNA that is present in all nucleated somatic cells. Over 90% of human DNA occurs in the nucleus, where it is distributed between the different chromosomes. The remaining DNA is found in mitochondria. Each human somatic cell nucleus contains 6×10^9 base pairs of DNA, which is equivalent to about 2 m of linear DNA. Packaging of the DNA is achieved by the double helix being coiled around histone proteins to form nucleosomes and then condensed further by coiling into the chromosome structure seen at metaphase. A single cell does not express all of its genes, and active genes are packaged into a more accessible chromatin configuration which allows them to be transcribed. Some genes are expressed at low levels in all cells and are called housekeeping genes. Others are tissue specific and are expressed only in certain tissues or cell types.

Chromosomes vary in size, containing between 60 and 263 megabases of DNA. Some chromosomes carry more genes than others, although this is not directly related to their size. Chromosomes 19 and 22, for example, are gene rich, whilst chromosomes 4 and 18 are gene poor. Many genes are members of gene families and have closely related sequences. These genes are often clustered, as with the globin gene clusters on chromosomes 11 and 16.

It is estimated that there are around 30 to 50 thousand pairs of functional genes in humans, yet these constitute only a small proportion of total genomic DNA. At least 95% of the genome consists of non-coding DNA (DNA that is not translated into a polypeptide product), whose function is not defined. Much of this DNA has a unique sequence, but between 30% and 40% consists of repetitive sequences that may be dispersed throughout the genome or arranged as regions of tandem repeats, known as satellite DNA. The repeat motif may consist of several thousand base pairs in megasatellites, 20–30 base pairs in minisatellites and simple 2 or 3 base pair repeats in microsatellties. In these tandem repeats the number of times that the core sequence is repeated varies among different people, giving rise to hypervariable regions. These are referred to as VNTRs (variable number of tandem repeats) and are stably inherited. Analysis of hypervariable minisatellite regions using a DNA probe for the common core sequence demonstrates DNA band patterns that are unique to a particular individual and this forms the basis of DNA fingerprinting tests.

Microsatellite repeats and other DNA variations due to differences in the nucleotide sequence that occur close to genes of interest can be used to track genes through families using DNA probes. This approach revolutionised the predictive tests available for mendelian disorders such as Duchenne muscular dystrophy and cystic fibrosis before the genes were isolated and the disease causing mutations identified.

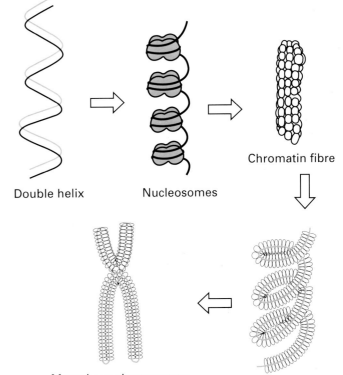

Chromatin fibre

Double helix   Nucleosomes

Metaphase chromosome   Loops of chromatin

Figure 15.8 Packaging of DNA into chromosomes

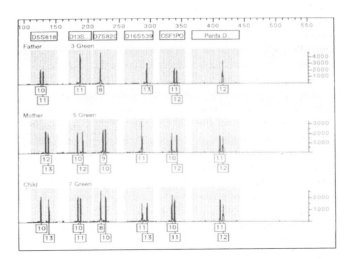

Figure 15.9 Fluorescent microsatellite analysis in a father (upper panel), mother (middle panel) and child (lower panel) for 5 markers. The marker name is indicated at the top of each set of traces. The child inherits one allele at each locus from each parent. (Data provided by Dr Andrew Wallace, Regional Genetic Service, St Mary's Hospital, Manchester)

16 Gene mapping and molecular pathology

International meetings on human gene mapping were inaugurated in 1973 and subsequently held every two years to document progress. At the first meeting the total number of autosomal genes whose chromosomal location had been identified was 64. The corresponding number of mapped genes had risen to 928 by the ninth meeting in 1987 as molecular techniques replaced those of traditional somatic cell genetics. The total number of mapped X linked loci also rose, from 155 in 1973 to 308 in 1987. The number of mapped genes has continued to increase rapidly since then, reflecting the development of new molecular biological techniques and the institution of the Human Genome Project.

Mendelian inheritance database

McKusick's definitive database, (Mendelian Inheritance in Man, Catalogs of Human Genes and Genetic Disorders. 12th edn Baltimore: Johns Hopkins University Press, 1998) has over the past 30 years, catalogued and cross-referenced published material on human inherited disorders, providing regular updates. The database has evolved in the face of an explosion of information on human genetics into a freely available on-line resource, which is being continually updated and revised.

The OMIM database (Online Mendelian Inheritance in Man) can be accessed via the US National Institute of Health website (www.ncbi.nih.nlm.gov/omim) or via the UK Human Gene Mapping Project Resource Centre website (www.hgmp.mrc.ac.uk/omim) and has over 12000 entries, summarised in the tables (OMIM Statistics for March 12, 2001).

Human Genome Project

The Human Genome Project was initiated in 1995 as an international collaborative project with the aim of determining the DNA sequence of each of the human chromosomes and of providing unrestricted public access to this information. Sequencing data have been submitted by 16 collaborating centres: eight from the United States, three from Germany, two from Japan and one from France, China, and the UK respectively. The UK contribution came from the Sanger Centre at Hinxton in Cambridgeshire, jointly funded by the Wellcome Trust and the Medical Research Council.

The human genome project consortium used a hierarchical shotgun approach in which overlapping bacterial clones were sequenced using mapping data from publicly available maps. Each bacterial clone was analysed to provide sequence data with 99.99% accuracy. The first draft of the human sequence covering 90% of the gene-rich regions of the human genome was published in a historic article in *Nature* in February 2001 (Volume 409, No 6822).

As a result of this monumental work, the overall size of the human genome has been determined to be 3.2 Gb (gigabases), making it 25 times larger that any genome previously sequenced. The consortium has estimated that there are approximately 32 000 human genes (far fewer than expected) of which 15 000 are known and 17 000 are predictions based on new sequence data.

The Human Genome Sequencing Project has been complicated by the involvement of commercial organisations. Celera Genomics started sequencing in 1998 using a whole genome shotgun cloning method and published its own draft

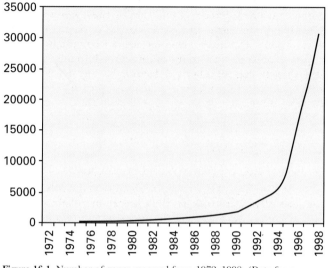

Figure 16.1 Number of genes mapped from 1972–1998. (Data from National Centre for Biotechnology Information, National Institute of Health, USA)

Table 16.1 Entries in the 'OMIM' database by mode of inheritance

OMIM entry	Autosomal	X linked	Y linked	Mito-chondrial	Total
Established genes or phenotype loci	8486	457	34	37	9014
Phenotypic descriptions	769	62	0	23	854
Other loci or phenotypes	2342	166	3	0	2511
Total	11 597	685	37	60	12 379

Table 16.2 Entries in the 'OMIM' database by chromosomal location

Chromosome	Loci	Chromosome	Loci	Chromosome	Loci
1	689	9	264	17	413
2	428	10	244	18	101
3	352	11	453	19	465
4	265	12	375	20	158
5	298	13	119	21	105
6	420	14	220	22	159
7	333	15	184	X	433
8	228	16	261	Y	30

Table 16.3 Progress in sequencing of human genome July 2001

	Total sequence (kb)	Percentage of genome (%)
Finished	1 660 078	47.10
Draft	3 547 899	51.40
Total	4 688 264	98.50

(Data from National Centre for Biotechnology Information, National Institute of Health, USA)

version of the human genome sequence in *Science* in February 2001 (Volume 291, No. 5507). Access to its information is restricted and Celera expect gene patent rights arising from use of its data.

Despite the huge milestone achieved by these human genome sequencing projects, the data generated represent only the first step in understanding the way genes work and interact with each other. The human genome sequence needs to be completed and coupled with further research into the molecular pathology of inherited diseases and the development of new treatments for conditions that are, at present, intractable.

Gene localisation

Prior to 1980, only a few genes, for disorders whose biochemical basis was known, had been identified. With the advent of molecular techniques the first step in isolating many genes for human diseases was to locate their chromosomal position by gene mapping studies. In some disorders, such as Huntington disease, this was achieved by undertaking linkage studies using polymorphic DNA markers in affected families, without any prior information about which chromosome carried the gene. In other disorders, the likely position of the gene was suggested by identification of a chromosomal rearrangement in an affected individual in whom it was likely that one of the chromosomal break points disrupted the gene. The neurofibromatosis type 1 (*NF1*) gene, for example, was isolated after the identification of such a translocation followed by cloning and sequencing of DNA from the region of the break point on chromosome 17.

In Duchenne muscular dystrophy, several affected females had been reported who had one X chromosome disrupted by an X:autosome translocation with the normal X chromosome being preferentially inactivated. The site of the break point in these cases was always on the short arm of the X chromosome at Xp21, which suggested that this was the location of the gene for DMD. DNA variations in this region, identified by hybridisation with DNA probes, provided markers that were shown to be linked to the gene for DMD in family studies in 1983. Strategies were then developed to identify DNA sequences from the region of the gene for DMD, some of which were missing in affected boys indicating that they represented deleted intragenic sequences. The entire gene for DMD was subsequently cloned in 1987 and its structure determined.

Gene tracking

Once a disease gene has been located using linkage analysis, DNA markers can be used to track the disease gene through families to predict the genetic state of individuals at risk. Prior to identifying specific gene mutations, this can provide information about carrier risk and enable prenatal diagnosis in certain situations. Before gene tracking can be used to provide a predictive test, family members known to be affected or unaffected must be tested to find an informative DNA marker within the family and to identify which allele is segregating with the disease gene in that particular kindred. Because recombination occurs between homologous chromosomes at meiosis, a DNA marker that is not very close to a gene on a particular chromosome will sometimes be inherited independently of the gene. The closer the marker is to a gene, the less likely it is that recombination will occur. In practice, markers that have shown less than 5% recombination with a

Table 16.4 Examples of mapped and cloned genes for each of the autosomes

Disorder	Chromosomal location	Gene symbol
Porphyria cutanea tarda	1p34	*UROD*
Waardenburg syndrome 1	2q35	*PAX3*
von Hippel–Lindau disease	3p26–p25	*VHL*
Huntington disease	4p16.3	*HD, IT15*
Familial adenomatous polyposis	5q21–q22	*APC*
Haemochromatosis	6p21.3	*HFE*
Cystic fibrosis	7q31.2	*CFTR*
Multiple exostoses 1	8p24	*EXT1*
Galactosaemia	9p13	*GALT*
Multiple endocrine neoplasia 2A	10q11.2	*RET*
Sickle cell anaemia and β-thalassaemia	11p15.5	*HHB*
Phenylketonuria (classical)	12q24.1	*PAH*
Wilson disease	13q14.3–q21.2	*ATP7B*
α₁ Antichymotrypsin deficiency	14q32.1	*AACT*
Tay–Sachs disease	15q23–q24	*HEXA*
Adult polycystic kidney disease 1	16p13.3–p13.2	*PKD1*
Neurofibromatosis 1 (peripheral)	17q11.2	*NF1*
Nieman–Pick type C	18q11–q12	*NPC1*
Familial hypercholesterolaemia	19p13.2	*LDLR*
Creutzfeldt-Jakob disease	20pter–p12	*PRNP*
Homocystinuria	21q22.3	*CBS*
Neurofibromatosis 2 (central)	22q12.2	*NF2*

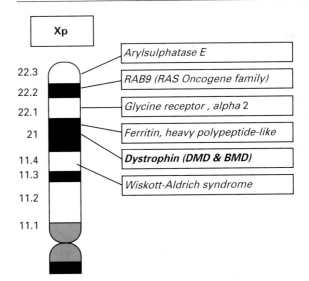

Figure 16.2 Short arm of chromosome X showing position of the dystrophin gene (mutated in Duchenne and Becker muscular dystrophies)

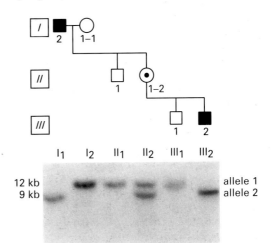

Figure 16.3 Tracking a DNA marker linked to the dystrophin gene through a family affected with Becker muscular dystrophy

disease gene have been useful in detecting carriers and in prenatal diagnosis, although there is always a margin of error with this type of test and results are quoted as a probability of carrying the gene and not as a definitive result. Linkage studies using intragenic markers provide much more accurate prediction of genetic state, but this approach is only used now when mutation analysis is not possible, as in some cases of Duchenne muscular dystrophy, Marfan syndrome and neurofibromatosis type 1.

Gene identification

Once the chromosomal location of a gene has been identified, there are several strategies that can be employed to isolate the gene itself. Genes within the region of interest can be searched for by using techniques such as cDNA selection and screening, CpG island identification and exon trapping. Any genes identified can then be studied for mutations in affected individuals. Alternatively, candidate genes can be identified by their function or expression patterns or by sequence homology with genes known to cause similar phenotypes in animals. The gene for Waardenburg syndrome, for example, was localised to chromosome 2q by linkage studies and the finding of a chromosomal abnormality in an affected subject. Identification of the gene was then aided by recognition of a similar phenotype in *splotch* mice. Mutations in the *PAX3* gene were found to underlie the phenotype in both mice and humans.

Types of mutation

In a few genetic diseases, all affected individuals have the same mutation. In sickle cell disease, for example, all mutant genes have a single base substitution, changing the sixth codon of the beta-globin gene from GAG to GTG, resulting in the substitution of valine for glutamic acid. In Huntington disease, all affected individuals have an expansion of a CAG trinucleotide repeat expansion. The majority of mendelian disorders are, however, due to many different mutations in a single gene. In some cases, one or more mutations are particularly frequent. In cystic fibrosis, for example, over 700 mutations have been described, but one particular mutation, ΔF508, accounts for about 70% of all cases in northern Europeans. In many conditions, the range of mutations observed is very variable. In DMD, for example, mutations include deletions, duplications and point mutations.

Deletions
Large gene deletions are the causal mutations in several disorders including α-thalassaemia, haemophilia A and Duchenne muscular dystrophy. In some cases the entire gene is deleted, as in α-thalassaemia; in others, there is only a partial gene deletion, as in Duchenne muscular dystrophy.

Duplications and insertions
Pathological duplication mutations are observed in some disorders. In Duchenne muscular dystrophy, 5–10% of mutations are due to duplication of exons within the dystrophin gene, and in Charcot–Marie–Tooth disease type 1a, 70% of mutations involve duplication of the entire *PMP22* gene. In DMD the mutation acts by causing a shift in the translation reading frame, and in CMT 1a by increasing the amount of gene product produced. Insertions of foreign DNA sequences into a gene also disrupt its function, as in haemophilia A caused by insertion of LINE1 repetitive sequences into the *F8C* gene.

Table 16.5 Notation of mutations and their effects

Notation of nucleotide changes

1657 G→T	G to T substitution at nucleotide 1657
1031–1032ins T	Insertion of T between nucleotides 1031 and 1032
1564delT	Deletion of a T nucleotide at nucleotide 1564
1063(GT)6–22	Variable length dinucleotide GT repeat unit at nucleotide 1063
IVS4–2A → T	A to T substitution 2 bases upsteam of intron 4
1997 + 1G →T	G to T substitution 1 base downstream of nucleotide 1997 in the cDNA

Notation of amino acid changes

Y92S	Tyrosine at codon 92 substituted by serine
R97X	Arginine at codon 97 substituted by a termination codon
T45del	Threonine at codon 45 is deleted
T97–98ins	Threonine inserted between codons 97 and 98 of the reference sequence

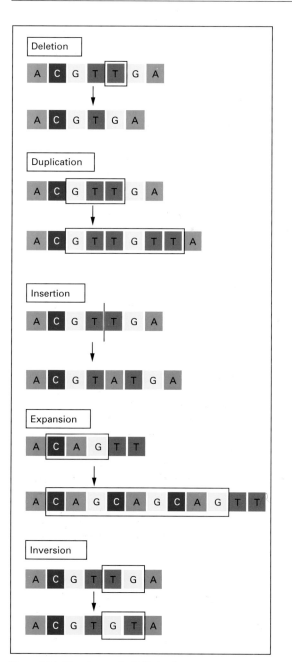

Figure 16.4 Mutation at the DNA level

Point mutations

Most disease-causing mutations are simple base substitutions, which can have variable effect. Mis-sense mutations result in the replacement of one amino acid with another in the protein product and have an effect when an essential amino acid is involved. Non-sense mutations result in replacement of an amino acid codon with a stop codon. This often results in mRNA instability, so that no protein product is produced. Other single base substitutions may alter the splicing of exons and introns, or affect sequences involved in regulating gene expression such as gene promoters or polyadenylation sites.

Frameshift mutations

Mutations that remove or add a number of bases that are not a multiple of three will result in an alteration of the transcription and translation reading frames. These mutations result in the translation of an abnormal protein from the site of the mutation onwards and almost always result in the generation of a premature stop codon. In Duchenne muscular dystrophy, most deletions alter the reading frame, leading to lack of production of a functional dystrophin protein and a severe phenotype. In Becker muscular dystrophy, most deletions maintain the correct reading frame, leading to the production of an internally truncated dystrophin protein that retains some function and results in a milder phenotype.

Trinucleotide repeat expansions

Expanded trinucleotide repeat regions represent new, unstable mutations that were identified in 1991. This type of mutation is the cause of several major genetic disorders, including fragile X syndrome, myotonic dystrophy, Huntington disease, spinocerebellar ataxia and Friedreich ataxia. In the normal copies of these genes the number of repeats of the trinucleotide sequence is variable. In affected individuals the number of repeats expands outside the normal range. In Huntington disease the expansion is small, involving a doubling of the number of repeats from 20–35 in the normal population to 40–80 in affected individuals. In fragile X syndrome and myotonic dystrophy the expansion may be very large, and the size of the expansion is often very unstable when transmitted from affected parent to child. Severity of these disorders correlates broadly with the size of the expansion: larger expansions causing more severe disease.

Epigenetic effects

Epigenetic effects are inherited molecular changes that do not alter DNA sequence. These can affect the expression of genes or the function of the protein product. Epigenetic effects include DNA methylation and alteration of chromatin configuration or protein conformation. Methylation of controlling elements silences gene expression as a normal event during development. Abnormalities of methylation may result in genetic disease. In fragile X syndrome, methylation of the promotor occurs when there is a large CGG expansion, inactivating the gene and causing the clinical phenotype. Methylation is also involved in the imprinting of certain genes, where abnormalities lead to disorders such as Angelman and Prader–Willi syndromes.

Modifier genes

The variation in phenotype between different affected members of the same family who have identical gene mutations may be due in part to environmental factors, but is probably also determined by the presence or absence of particular alleles at other loci, referred to as modifier genes. Modifying genes may for example, determine the incidence of complications in

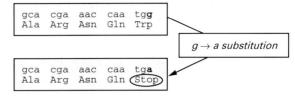

Non-sense mutation

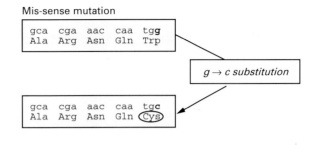

Mis-sense mutation

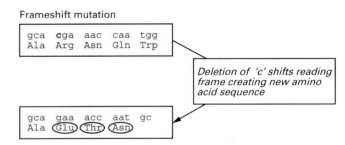

Frameshift mutation

Figure 16.5 Effect of mutations at the amino acid level

Box 16.1 Properties of trinucleotide repeat regions

- Trinucleotide repeat numbers in the normal range are stably inherited and have no adverse phenotypic effect
- Trinucleotide repeat numbers outside the normal range are unstable and may expand further when transmitted to offspring
- Adverse phenotypic effects occur when the size of the expansion exceeds a critical length

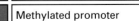
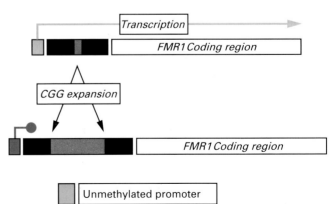
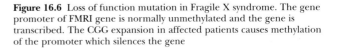

Figure 16.6 Loss of function mutation in Fragile X syndrome. The gene promoter of FMRI gene is normally unmethylated and the gene is transcribed. The CGG expansion in affected patients causes methylation of the promoter which silences the gene

insulin dependent diabetes, the development of amyloidosis in familial Mediterranean fever and the occurrence of meconium ileus in cystic fibrosis.

Abnormalities of gene function

Different types of genetic mutation have different consequences for gene function. The effects on phenotype may reflect either loss or gain of function. In some genes, either type of mutation may occur, resulting in different phenotypes.

Loss of function mutations

Loss of function mutations result in reduced or absent function of the gene product. This type of mutation is the most common, and generally results in a recessive phenotype, in which heterozygotes with 50% of normal gene activity are unaffected, and only homozygotes with complete loss of function are clinically affected. Occasionally, loss of function mutations may have a dominant effect. Heterozygosity for chromosomal deletions usually causes an abnormal phenotype and this is probably due to haploinsufficiency of a number of genes.

Many different mutation types can result in loss of function of the gene product and when a variety of mutations in a gene cause a single phenotype, these are all likely to represent loss of function mutations. In fragile X syndrome, for example, the most common mutation is a pathological expansion of a CGG trinucleotide repeat that silences the *FMR1* gene. Occasionally the syndrome is due to a point mutation in the *FMR1* gene, also associated with lack of the gene product that produces the same phenotype.

Dominant negative effect

In some conditions, the abnormal gene product not only loses normal function but also interferes with the function of the product from the normal allele. This type of mutation acts in a dominant fashion and is referred to as having a dominant negative effect. In type I osteogenesis imperfecta (OI), for example, the causal mutations in the *COL1A1* and *COL1A2* genes produce an abnormal type I collagen that interferes with normal triple helix formation, resulting in production of an abnormal mature collagen responsible for the OI phenotype.

Gain of function mutation

When the protein product produced by a mutant gene acquires a completely novel function, the mutation is referred to as having a gain of function effect. These mutations usually result in dominant phenotypes because of the independent action of the gene product. The CAG repeat expansions in Huntington disease and the spinocerebellar ataxias exert a gain of function effect, by resulting in the incorporation of elongated polyglutamine tracts in the protein products. This causes formation of intracellular aggregates that result in neuronal cell death. Mutations producing a gain of function effect are likely to be very specific and other mutations in the same gene are unlikely to produce the same phenotype. In the androgen receptor gene, for example, a trinucleotide repeat expansion mutation results in the phenotype of spinobulbar muscular atrophy (Kennedy syndrome), whereas a point mutation leading to loss of function results in the completely different phenotype of testicular feminisation syndrome.

Overexpression

Overexpression of a structurally normal gene may occasionally produce an abnormal phenotype. Complete duplication of the

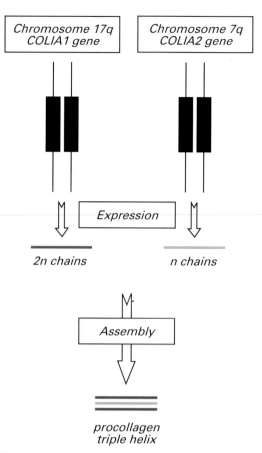

Figure 16.7 Mutations in genes involved in the synthesis of multimeric proteins such as collagens are prone to 'dominant negative' effects as the protein relies on the normal expression of more than one gene

---

Box 16.2 Examples of disorders caused by CAG repeat expansions conferring a gain of function

- Huntington disease
- Kennedy syndrome (SBMA)
- Spinocerebellar ataxias   SCA 1
                             SCA 2
                             SCA 6
                             SCA 7
- Machado–Joseph disease   SCA 3
- Dentatorubro–Pallidolysian atrophy (DRPLA)

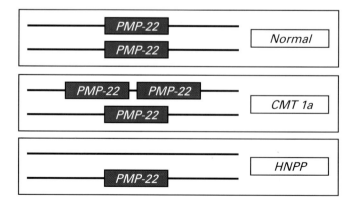

Figure 16.8 In Charcot–Marie–Tooth disease, the commonest form (Clinical type 1a) is caused by 1.5 Mb duplication that creates an extra copy of the *PMP22* gene. The milder HNPP is caused by deletion of one copy of the *PMP22* genes

PMP22 gene, with an increase in gene product, results in Charcot–Marie–Tooth disease type 1a. Interestingly, point mutations in the same gene produce a similar phenotype by functioning as activating mutations. Although examples of

gene duplication are not common, the abnormal phenotype associated with chromosomal duplications is probably due to the overexpression of a number of genes.

17   Techniques of DNA analysis

With the huge increase in knowledge of the human genome and its DNA sequence, growing numbers of disease genes can now be examined using DNA analysis. Few laboratory tests at the disposal of the modern clinician have the potential specificity and information content of these techniques. Only a few years ago, DNA analysis was mainly applicable to presymptomatic diagnosis of inherited conditions and the detection of carriers following initial diagnosis of the patient by more conventional laboratory tests (e.g. biochemical and histological). In current practice, the DNA laboratory has an increasing role in the initial diagnosis of many diseases by analysis of specific genes associated with mendelian disorders.

Over 20 regional molecular genetics laboratories provide a service to the regions of the UK with many additional laboratories providing genetic tests in areas such as mitochondrial disease and haemoglobinopathies. The following chapter summarises the standard techniques of DNA analysis employed by molecular laboratories for the provision of services to the clinician.

DNA extraction

Genomic DNA is usually isolated from EDTA-anticoagulated whole blood, often using an automated method. In addition, DNA can also be readily isolated from fresh or frozen tissue samples, chorionic villus biopsies, cultured amniocytes and lymphoblastoid cell lines. Smaller quantities of DNA can be recovered from buccal mouthwash samples and fixed embedded tissues, although the recovery is considerably less reliable. The increased use of the polymerase chain reaction (PCR) means that for a small proportion of analyses, blood volumes of <1 ml are adequate. In many instances however, larger volumes of blood are still required because numerous tests are required when analysing large or multiple genes and not all tests use PCR based methods of analysis.

Genomic DNA remains stable for many years when frozen. This enables storage of samples for future analysis of genes that are not yet isolated, and is crucial when organising the collection of DNA samples for long term studies of inherited conditions.

The polymerase chain reaction (PCR)

The use of PCR in the analysis of an inherited condition was first demonstrated in the detection of a common β-globin mutation in 1985. Since then, PCR has become an indispensable technique for all laboratories involved in DNA analysis. The technique requires the DNA sequence in the gene or region of interest to have been elucidated. This limitation is becoming increasingly less problematic with the pending completion of the entire human DNA sequence.

The main advantage of the PCR method is that the regions of the gene of interest can be amplified rapidly using very small quantities of the original DNA sample. This feature makes the method applicable in prenatal diagnosis using chorionic villus or amniocentesis samples and in other situations in which blood sampling is not appropriate.

The first step in PCR is to heat denature the DNA into its two single strands. Two specific oligonucleotide primers (short

Figure 17.1 Clinical scientist carrying out DNA sequencing analysis

Figure 17.2 Blood samples undergoing lysis during DNA extraction. As little as 30 μl of whole blood can provide sufficient DNA for a simple PCR-based analysis

Figure 17.3 Automated instrument for the extraction of DNA from blood samples of 5–20 ml volumes

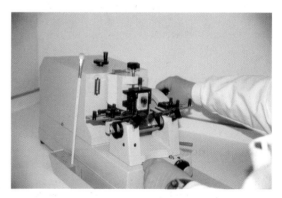

Figure 17.4 DNA extracted from paraffin-embedded pathology blocks may be useful in analysis of previous familial cases of conditions such as inherited breast cancer

synthetic DNA molecules), which flank the region of interest, are then annealed to their complementary strands. In the presence of thermostable polymerase, these primers initiate the synthesis of new DNA strands. The cycle of denaturation, annealing, and synthesis repeated 30 times will amplify the DNA from the region of interest 100 000-fold, whilst the quantity of other DNA sequences is unchanged.

In practice, because of the way genomic DNA is organised into coding sequences (exons) separated by non-coding sequences (introns), analysis of even a small gene usually involves multiple PCR amplifications. For example, the breast cancer susceptibility gene, *BRCA1*, is organised into 24 exons, with mutations potentially located in any one of them. Analysis of *BRCA1* therefore necessitates PCR amplification of each exon to enable mutation analysis.

Figure 17.5 DNA thermal cyclers used for PCR amplification of DNA

Post-PCR analysis

It should be noted that the PCR process itself is usually merely a starting point for an investigation by providing a sufficient quantity of DNA for further analysis. After completion of thermal cycling, the first step in analysis is to determine the success of amplification using agarose gel electrophoresis (AGE). The DNA is separated within the gel depending on its size; large DNA molecules travel slowly through the gel in contrast to small DNA molecules that travel faster. The DNA is detected within the gel with the use of a fluorescent dye (ethidium bromide) as a pink fluoresent band when illuminated by ultraviolet light. By varying the agarose concentration in the gel, this approach can be used for the analysis of PCR products from less than 100 to over 10 000 base pairs in size.

As well as showing the presence or absence of a PCR product, an agarose gel can also be used to determine the size of the product. In some instances, agarose gel electrophoresis alone is sufficient to demonstrate that a mutation is present. For example, a 250 base pair PCR product containing a deletion mutation of 10 bases will be readily detected by agarose gel electrophoresis. Determining the exact position of the deletion, however, requires additional analysis.

Agarose gel electrophoresis is of sufficient resolution to allow the rapid detection of the deletion of whole exons, which is often seen in affected male DMD patients. In this approach, a number of exons of the DMD gene are simultaneously amplified in a "multiplex" PCR approach. Samples with exon deletions are readily detected by the absence of specific bands when analysed by agarose gel electrophoresis.

For analysis of PCR products below 1000 bp, polyacrylamide gel electrophoresis is often used, which allows separation of DNA molecules that differ from each other in size by only a single base. The DNA can be detected in the gel by a variety of methods including ethidium bromide staining and silver staining however, many laboratories now use fluorescently tagged primers to generate labelled PCR products that can be visualised by laser-induced fluorescence. It is this technology that has been developed into the high-throughput DNA sequencing instruments that have been the workhorses of the Human Genome Sequencing Project.

Sequence-specific amplification
One of the properties of the short synthetic pieces of DNA (oligonucleotides) used as primers in PCR is their sequence specificity. This can be exploited to design PCR primers that only generate a product when they are perfectly matched to their target sequence. Conversely, a mismatch in the region of

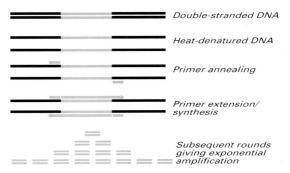

Double-stranded DNA

Heat-denatured DNA

Primer annealing

Primer extension/ synthesis

Subsequent rounds giving exponential amplification

Figure 17.6 Diagrammatic representation of PCR

Figure 17.7 PCR amplified DNA being loaded onto an agarose gel before electrophoresis

Figure 17.8 Visualisation of amplified DNA by ultra-violet transillumination. The DNA can be seen as pink/orange bands on the illuminated gel

sequence where the primer binds, prevents PCR amplification from proceeding. In this way, an assay can be designed to detect the presence or absence of specific known mutations. This approach (known as 'ARMS' or Amplification Refractory Mutation System) is often used to detect common cystic fibrosis mutations and certain mutations involved in familial breast cancer.

Oligonucleotide ligation assay (OLA)

In the OLA reaction, two oligonucleotide probes are hybridised to a DNA sample so that the 3′ terminus of the upstream oligo is adjacent to the 5′ terminus of the downstream oligo. If the 3′ terminus of the first primer is perfectly matched to its target sequence, then the probes can be joined together with a DNA ligase. In contrast no ligation can occur if there is a mismatch at the 3′ terminus of the first oligo. This approach has been successfully applied to the detection of 31 common mutations in cystic fibrosis with a commercial kit, and for the detection of 19 common mutations in the LDL receptor gene in hypercholesterolaemia.

Restriction enzyme analysis of PCR products

Restriction endonuclease enzymes are produced naturally by bacterial species as a mechanism of protection against "foreign" DNA. Each enzyme recognises a specific DNA sequence and cleaves double-stranded DNA at this site. Hundreds of these restriction enzymes are now commercially available and provide a rapid and reliable method of detecting the presence of a specific DNA sequence within PCR products. This property becomes especially relevant when a mutation either creates or destroys the enzyme's recognition site. By studying the size of the products that are generated following restriction enzyme digestion of PCR-amplified DNA (by agarose gel electrophoresis), it is possible to accurately determine the presence or absence of a particular mutation.

Single-stranded conformation polymorphism analysis (SSCP)

The principle of SSCP analysis is based on the fact that the secondary structure of single-stranded DNA is dependent on its base composition. Any change to the base composition introduced by a mutation or polymorphism will cause a modification to the secondary structure of the DNA strand. This altered conformation affects its migration through a non-denaturing polyacrylamide gel, resulting in a band shift when compared to a sample without a mutation. The bands of single-stranded DNA are usually visualised by silver-staining. It should be noted that the presence of a band shift itself does not provide any information about the nature of the mutation. Consequently, samples that show altered banding patterns require further investigation by DNA sequencing.

Heteroduplex analysis

Heteroduplexes are double-stranded DNA molecules that are formed from two complementary strands that are imperfectly matched. If a mutation is present in one copy of a gene being amplified using PCR, heteroduplexes will be formed from the hybridisation of the normal and the mutant PCR product. As in SSCP analysis described above, these structures will have altered mobility when analysed through non-denaturing polyacrylamide gels, and are seen as band shifts when compared to perfectly matched PCR products (or homoduplexes).

In practice, SSCP and heteroduplex analysis can be carried out simultaneously on the same polyacrylamide gel to increase the sensitivity of the analysis.

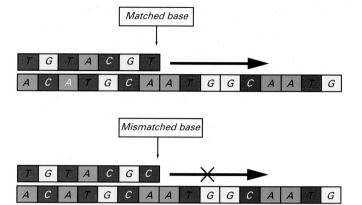

Figure 17.9 Sequence-specific PCR. For an oligonucleotide to act as a primer in PCR the 3' end (i.e. the end that it extends from) must be perfectly matched with its template. This property can be exploited to design a test that interrogates a specific DNA base (e.g. for detection of common breast cancer mutations)

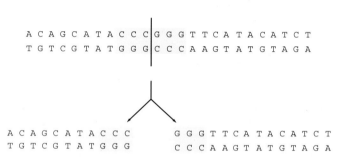

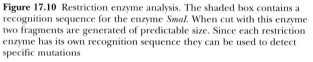

Figure 17.10 Restriction enzyme analysis. The shaded box contains a recognition sequence for the enzyme *SmaI*. When cut with this enzyme two fragments are generated of predictable size. Since each restriction enzyme has its own recognition sequence they can be used to detect specific mutations

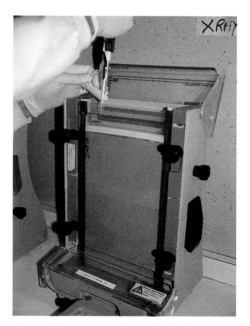

Figure 17.11 Loading PCR-amplified DNA onto an SSCP/heteroduplex gel

Denaturing gradient gel electrophoresis (DGGE)

The DGGE method relies on the fact that double-stranded DNA molecules have specific denaturation characteristics, i.e. conditions at which the double-stranded DNA disassociates into its two single-stranded units. The denaturation of the DNA strands can be achieved by increasing temperature or by the addition of a chemical denaturant such as urea or formamide. If a PCR product contains a mutation, this will subtly modify the conditions at which denaturation occurs, which in turn affects its electrophoretic mobility. In DGGE, a gradient of the denaturing agent is set up so that the PCR products migrate through the denaturant and are separated based on their sequence specific mobility.

Denaturing HPLC (DHPLC)

While conventional SSCP and heteroduplex analysis use polyacrylamide gel electrophoresis to separate PCR products, DHPLC uses a high pressure system to force the products through a column under partially denaturing conditions. Conditions for optimum separation of normal and mutant sequences are created by the use of buffer gradients and specific temperatures. The DNA molecules that are progressively eluted from the column are monitored by an ultraviolet detector with data being collected by computer.

Protein truncation test (PTT)

The key features of PTT are (i) that the analysis is based on the protein product generated from the DNA sequence, and (ii) the method specifically detects premature protein truncation caused by non-sense mutations. The PCR product is transcribed and translated in vitro by a reticulocyte lysate, during which the nascent protein product is radiolabelled with ^{35}S-labelled amino acids. The translation products are then separated by polyacrylamide gel electrophoresis. Samples with non-sense mutations are detected by their tendency to generate smaller protein products than their normal counterparts.

Chemical and enzymatic cleavage of mismatch (CCM)

As outlined in previous sections, PCR products that contain point mutations form hybrid molecules with their normal counterparts known as heteroduplexes. The two DNA strands in these heteroduplexes are perfectly matched except at the site of the mutation, where base pairing cannot occur. These mismatched sites can be recognised both by specific enzymes and by chemicals such as osmium tetroxide and piperidine, which cleave the DNA at the site of mismatch. This property can therefore be used to detect mutations within a PCR product by polyacrylamide gel electrophoresis to visualise the cleavage products.

DNA sequencing

In many of the techniques outlined above, no specific information is gained about the exact nature of the alteration in the DNA. In some cases, the change detected may turn out to be a polymorphism that has no direct bearing on the condition under investigation. The exception to this is the protein truncation test (PTT), which detects mutations that shorten the protein product and are therefore more likely to be pathogenic. In chemical cleavage of mismatch analysis, particular types of base mismatch are cleaved specifically by the different chemicals employed; this yields limited information about the type of change observed.

However, to determine the precise nature of the structure of the gene under investigation, DNA sequencing must be carried out. The commonest type of DNA sequencing in use

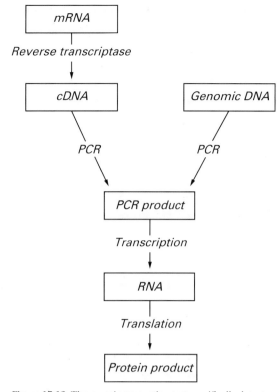

Figure 17.12 The protein truncation test specifically detects mutations that result in in-vitro premature translation termination

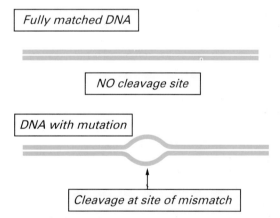

Figure 17.13 Naturally-occurring enzymes involved in DNA repair can be used to detect mutations since they cut double-stranded DNA at regions of mismatch. The same effect can also be created using chemical methods

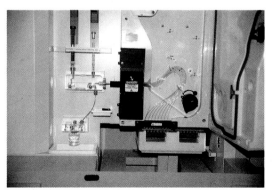

Figure 17.14 Interior of third-generation automated sequencing instrument in which DNA molecules are separated through fine capillaries

today (so called dideoxy or chain terminating) was invented by Fred Sanger in 1977. The technique was further refined using technology developed prior to the Human Genome Project and is now a routine method of analysis in many molecular genetic laboratories.

The technique relies on making a copy of the DNA in the presence of modified versions of the four bases (A, C, G, and T) which are fluorescently labelled with their own specific tag. The sequencing products are then separated with the use of long polyacrylamide gels with a laser being used to automatically detect the fluorescent molecules as they migrate. A computer program is then used to generate the DNA sequence. Recent improvements in DNA sequencing have seen polyacrylamide gels being replaced by capillary columns allowing the method to be further automated.

Hybridisation methods and "gene-chip" technology

In most of the methods described above, the specific site of a mutation within a gene is not known until after DNA sequencing has been completed. If the mutation is very common, however, methods may be used that specifically interrogate the site of the mutation. One of the simplest ways of doing this is by using a restriction enzyme (see above); however, this is not applicable in all situations.

Another possibility is the use of DNA probe technology. This utilises the tendency of two complementary single-stranded DNA molecules to anneal together to produce a double-stranded duplex. This method involves the DNA under investigation being immobilised onto a solid support such as nylon. A labelled single-stranded DNA probe may then be used to determine whether a specific sequence is present. This technique is often referred to as forward dot-blotting.

Alternatively, the probes may be immobilised to the membrane and hybridised with the labelled target DNA, that is free in solution (the reverse dot-blot approach). It is this basic principle that has been developed into the so-called "gene chip" technology. In this technique, literally thousands of short DNA probe molecules are first attached to silica-based support materials. The DNA under investigation is then fluorescently labelled and hybridised to the probe matrix. The large number of probes used enables the pattern of hybridisation to be translated into sequence information. At present, however, the high cost of this approach means that it is of limited value for the analysis of rare disease genes in a diagnostic setting.

Non-PCR based analysis

Not every gene can be studied using PCR. In some conditions, the mutation itself is large, and may have even deleted the entire gene. In other cases, the gene may be very rich in G and C bases, which makes conventional PCR difficult. In these situations, the older methods of analysis are invaluable, although generally more time-consuming than PCR-based methods.

Southern blotting

Although largely replaced by PCR-based methods, Southern blotting is still necessary to detect relatively large changes in the DNA that exceed the limits of PCR. Genomic DNA is first cut using restriction enzymes and the digested fragments fractionated using gel electrophoresis. The DNA is then transferred by capillary blotting onto nylon membrane before radiolabelled probes are used to investigate the region of interest.

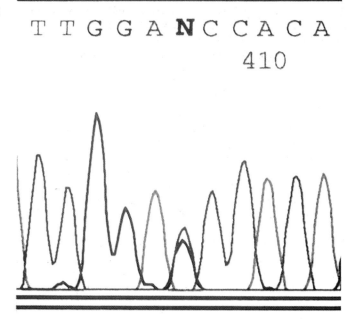

T T G G A **N** C C A C A
410

Figure 17.15 Output from DNA sequencer showing single nucleotide substitution, detected by the analysis software as an 'N'

Figure 17.16 Affymetrix GeneChip® probe array (courtesy of Affymetrix)

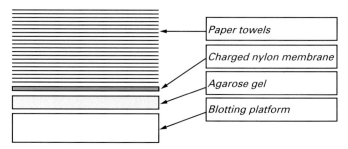

Paper towels

Charged nylon membrane

Agarose gel

Blotting platform

Figure 17.17 Setting-up a Southern blot (dry-blotting). Using a stack of paper towels to provide capillarity, the DNA in the agarose gel is transferred to the charged membrane before being hybridised with a radiolabelled DNA probe

Pulse-field gel electrophoresis (PFGE)
In a development of standard Southern blotting methods, PFGE uses specialised restriction enzymes and electrophoresis conditions to fractionate the genomic DNA to a high-resolution. This method is more applicable to the detection of large deletions, well out of the range of PCR.

Future developments

DNA sequencing currently provides information on the order of bases within a gene to a high degree of accuracy. However, the large size of many genes involved (e.g. the breast cancer susceptibility genes *BRCA*1 and *BRCA*2) and the number of patients requiring analysis means that improvements in throughput are highly desirable. Robotic workstations are currently being introduced into many molecular genetic laboratories to try to meet this demand by automating many of the laborious sample handling steps involved.

In addition to improvements in sample throughput, molecular genetic laboratories are increasingly paying attention to the functional significance of the genetic changes that they detect. Functional studies are especially important in predictive and pre-symptomatic analysis, where the relevance of a mutation has a direct bearing on the decision making process. The vast quantity of information that has been generated by the Human Genome Project will undoubtedly increase the ability to predict the effect of specific mutations. However, there may well come a time when the detection of a genetic event is only the first stage in the investigation into its functional effect.

Figure 17.18 Pulse-field gel electrophoresis (PFGE) equipment. In this technique an electric current is passed through the gel in timed pulses at differing angles to separate very large DNA molecules. Note the hexagonal arrangement of the electrodes in this case

18 Molecular analysis of mendelian disorders

Molecular genetic analysis is now possible for an increasing number of single gene disorders. In some cases direct mutation detection is feasible and molecular testing will provide or confirm the diagnosis in the index case in a family. This enables tests to be offered to other relatives to provide presymptomatic diagnosis, carrier testing and prenatal diagnosis as appropriate. For recessive conditions that are due to a small number of gene mutations, or those that have a commonly occurring mutation, it may also be possible to offer molecular based carrier tests to an unrelated spouse. Tests for very rare disorders in the UK are usually carried out on a national basis by designated laboratories. For the more common disorders, genetic analysis is undertaken in most of the regionally based NHS molecular genetic laboratories. In this chapter, examples of some of these common inherited disorders have been chosen to illustrate the range of tests performed.

Haemoglobinopathies

The haemoglobinopathies are a heterogeneous group of inherited disorders characterised by the absent, reduced or altered expression of one or more of the globin chains of haemoglobin. The globin gene clusters on chromosome 16 include two α-globin genes and on chromosome 11 a β-globin gene. The haemoglobinopathies represent the commonest single-gene disorders in the world population and have had profound effects on the provision of health care in some developing countries.

Various mutations in the β-globin gene cause structural alterations in haemoglobin, the most important being the point mutation that produces haemoglobin S and causes sickle cell disease. Direct detection of this point mutation permits carrier detection and first-trimester prenatal diagnosis.

The thalassaemias are due to a reduced rate of synthesis of α- or β-globin chains, leading to an imbalance in their production. α-thalassaemia is a defect of α-globin chain synthesis. Each normal adult chromosome expresses two copies of the α-globin gene and disease severity is proportional to the number of α-globin genes lost following a mutational event. In the most severe type, Barts hydrops fetalis, all four copies are lost, leading to a severe phenotype associated with stillbirth or early neonatal death. The α-globin gene cluster contains a number of repeat regions that increase the likelihood of unequal crossover during meiosis. As a result, relatively large deletions are the commonest type of mutations that give rise to α-thalassamia. In particular, a 3.7 kilobase (kb) deletion is common in patients from Africa, the Mediterranean, Middle East and India. A 4.2 kb deletion is common in patients from southeast Asia and the Pacific Islands. Both 3.7 kb and 4.2 kb deletions can be detected by PCR analysis; however, since amplification of the region is often technically challenging, Southern blotting is still considered a reliable method of analysis.

β thalassaemia results from a variety of molecular defects that either reduce or completely abolish β-globin synthesis. Over 200 mutations have so far been reported with point mutations and small deletions comprising the majority. Although a large number of mutations have been reported, the prevalence of specific mutations is dependent on the ethnic origin. Diagnostic testing therefore requires knowledge of the mutation spectrum in the population being screened. Eighty per cent of mutations

Figure 18.1 Sites of Regional Molecular Genetics Laboratories in the UK and Ireland

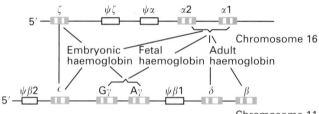

Figure 18.2 Globin gene clusters on chromosomes 11 and 16 (ψ denotes pseudogene)

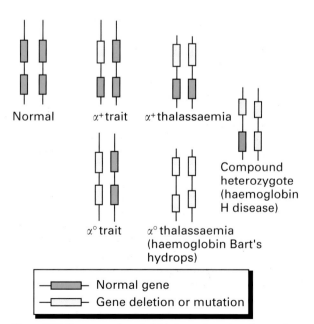

Figure 18.3 Representation of globin genes in various forms of α-thalassaemia

can be detected in a reverse dot-blot approach in which PCR-amplified DNA is labelled and hybridised against a panel of probes immobilised onto a nylon strip. Alternatively an ARMS PCR approach may be used, which is especially useful for rapid mutation analysis in a prenatal setting.

Cystic fibrosis (CF)

Along with fragile X syndrome, CF represents the commonest request for analysis to most molecular genetic laboratories, because of the high frequency of carriers in the population (1 in 22 in the UK). The incidence of CF varies between approximately 1 in 2000 live births in white caucasians to 1 in 90 000 in asians.

Cystic fibrosis is caused by mutations in the cystic fibrosis transmembrane regulator (*CFTR*) gene located on the long arm of chromosome 7, which contains 27 exons. Approximately 700 mutations have been described, many of which are "private" mutations restricted to a particular lineage. Approximately half the mutations are "mis-sense" (i.e. the protein product is full length but contains an amino acid substitution). The commonest single mutation in CF is a deletion known as ΔF508 that accounts for at least 70% of cystic fibrosis mutations in northern Europeans.

Most molecular genetic laboratories will test for the commonest CF mutations using either an ARMS (amplification refractory mutation system) PCR analysis or other mutation-specific tests such as OLA (oligonucleotide ligation assay). It should be remembered that since the frequency of mutations varies between populations, the panel of mutations tested in one ethnic group may be of less value in another ethnic group and consequently knowledge of the mutation spectrum in the local population is important.

Fragile X syndrome (FRAX–A)

Fragile X syndrome is one of a group of disorders caused by the expansion of a triplet repeat region within a gene. It is associated with the presence of a fragile site on the X chromosome (Xq27.3), categorised as FRAX-A. The syndrome is characterised by mental retardation and accounts for 15–20% of all X linked mental retardation. Affected males have moderate to severe mental retardation, whereas affected females have milder retardation and phenotypic features.

Fragile X syndrome is caused by an expanded CGG repeat in the untranslated region of the *FMR-1* gene, which results in reduction or abolition of expression of the gene by methylation of the gene promoter. In normal individuals, the number of CGG repeats varies between 6 and 54 units and is stably inherited. However, if individuals have between 55 and 200 repeats (although apparently unaffected), there is an increased risk of the repeat region expanding further into the full mutation range (>200 repeats) that is associated with mental retardation.

The fragile site associated with FRAX–A may be detected using cytogenetic methods by culturing cells in the absence of folic acid and thymidine but this is not a sensitive test for detecting carrier females. The expansion of the CGG repeat in the *FMR-1* gene may be detected at the DNA level using PCR. After amplification, the size of the repeat from each chromosomal copy is determined by polyacrylamide gel electrophoresis. Samples with a known number of repeats are used as size standards. This type of approach can be used only as a screen to detect normal sized alleles. Because full mutations with long stretches of CGG repeats are too large to amplify effectively, Southern blotting is still widely used in

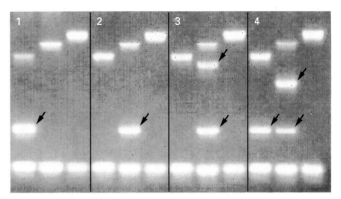

Figure 18.4 Detection of known mutations in the *CFTR* gene using the Elucigene™ CF20 mutation kit (Orchid Biosciences, Abingdon, UK). Mutations are detected by the presence of specific bands in an agarose gel.
Sample 1: ΔF508 homozygote
Sample 2: Normal pattern
Sample 3: 621+1g→t heterozygote
Sample 4: ΔF508, R117H compound heterozygote
(courtesy of Dr Simon Ramsden, Regional Genetic Service, St. Mary's Hospital, Manchester)

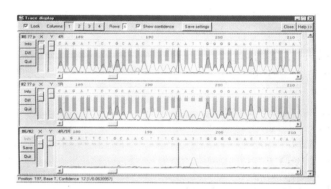

Figure 18.5 Semi-automated detection of mutations such as *CFTR* using the Gap4 sequence analysis software. (Bonfield et al., Nucleic Acids Research 14, 3404–3409, 1998) The algorithm subtracts the trace of a control sample from the trace of a test sample, highlighting mutations and polymorphisms (see lower panel) (screen shot courtesy of Dr Karen Young, Regional Genetic Service, St. Mary's Hospital, Manchester)

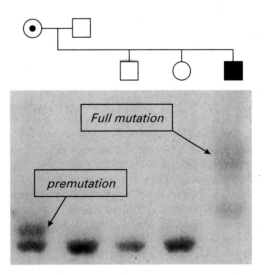

Figure 18.6 Southern blot analysis of the trinucleotide repeat region in the *FMR-1* gene associated with fragile X syndrome (FRAX-A) (courtesy of Dr Simon Ramsden, Regional Genetic Service, St. Mary's Hospital, Manchester)

FRAX–A analysis. This method can also be modified to determine the methylation status of the gene (the main influence on normal *FMR-1* gene expression). In prenatal diagnosis, methylation analysis is problematic owing to the presence of fetal methylation patterns, and the size of the repeat becomes the most reliable predictive indicator.

Huntington disease (HD)

HD is a progressive fatal neurodegenerative disease. Like FRAX–A, HD is caused by a triplet repeat expansion. The HD expansion involves a CAG triplet in exon 1 of the *IT15* gene on chromosome 4. The expansion is translated into a polyglutamine tract in the huntingtin protein gene product that is believed to cause a dominant gain of function leading to neuronal loss.

In normal individuals, the CAG unit in exon 1 has between 9 and 35 repeats. Affected individuals have repeats of 36 units or greater, with over 90% of affected subjects having 40–55 repeats. In general, the greater the number of repeats an individual has, the earlier the age of onset will be, although this relationship is stronger for higher repeat numbers.

Since the CAG repeat expansion is the sole mutation responsible for all HD cases, molecular genetic analysis concentrates on this single region. Small CAG expansions can be detected using PCR amplification of the repeat region. The PCR products are then sized using polyacrylamide gel electrophoresis. Samples with known repeat sizes may be used as controls to determine the size of the expansion. Larger expansions cannot be detected by PCR and the time-consuming Southern blotting method must be used in cases where two normal sized repeat alleles are not detected by PCR.

Charcot–Marie–Tooth disease (CMT)

CMT disease (or hereditary motor and sensory neuropathy, HMSN) is clinically and genetically heterogeneous, but is generally characterised by wasting and weakness of the distal limb muscles with or without distal sensory loss. CMT may be inherited in an autosomal dominant, autosomal recessive or X linked manner. Clinically, the condition is divided into the demyelinating CMT1 (with reduced nerve conduction velocities) and axonal CMT2 (with nerve conduction velocites largely preserved). Rarer clinical forms exist, including the severe Dejerine–Sottas syndrome and hereditary neuropathy with increased reflexes. The related condition HNPP (hereditary neuropathy with liability to pressure palsies) creates a milder phenotype characterised by recurrent, usually transient sensorimotor neuropathies.

Many of the genes involved in CMT have now been cloned and sequenced, allowing a genetic classification to be made depending on the mutation or gene locus identified. Mutations in over five genes have been reported in CMT, including *PMP22* (peripheral myelin protein on chromosome 17), *MPZ* (myelin protein zero on chromosome 1), *Connexin-32* (X chromosome), *EGR2* and *NEFL*. The commonest mutational event is the duplication of the entire *PMP22* gene resulting in clinical CMT type 1a. A deletion of the same gene gives rise to the milder HNPP phenotype. Phenotypes of varying severity can also be produced by point mutations (often base substitutions) in any of the five genes mentioned above. Prediction of disease severity in presymptomatic patients is difficult as there is varying severity even within families.

Detection of the duplication or deletion of the *PMP22* gene is achieved using fluorescent dosage PCR analysis to determine

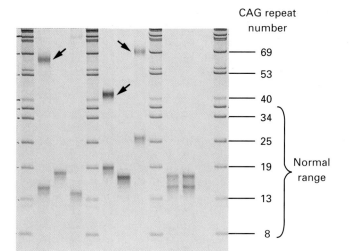

Figure 18.7 PCR analysis of the trinucleotide repeat region associated with Huntington disease. Note that a number of the samples have repeat expansions within the pathological range as indicated by the arrows (courtesy of Alan Dodge, Regional Genetic Service, St. Mary's Hospital, Manchester)

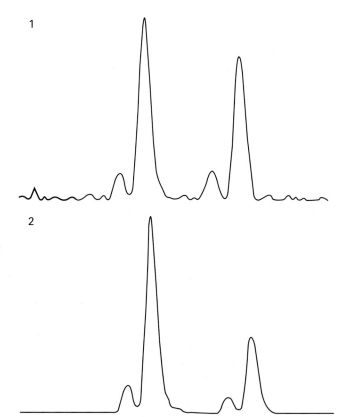

Figure 18.8 Detection of the 17p11.2 duplication associated with CMT/HMSN type 1a by fluorescent dosage analysis using flanking microsatellite markers
Panel 1: Duplication-negative sample with approximately equal contributions from each allele.
Panel 2: Duplication-positive sample with a 2:1 ratio of alleles (courtesy of Dr David Gokhale, Regional Genetic Service. St. Mary's Hospital, Manchester)

the number of gene copies present. Following initial PCR amplification with fluorescently-labelled primers, the products are analysed by automated laser-induced fluorescence. Point mutations in all five CMT genes are detected by a variety of methods depending on local practices, including SSCP, DGGE and DNA sequencing. Requests for prenatal testing in the UK are rare.

Spinal muscular atrophy (SMA)

SMA encompasses a clinically and genetically heterogeneous group of disorders characterised by degeneration and loss of the anterior horn cells in the spinal cord and sometimes in the brainstem nuclei, resulting in muscle weakness and atrophy. Most cases are inherited in an autosomal recessive fashion, although some affected families show dominant inheritance. Childhood onset SMA is the second most common, lethal autosomal recessive disorder in white populations, with an overall incidence of 1 in 10 000 live births and a carrier frequency of approximately 1 in 50. It is estimated to be the second most frequent disease seen in paediatric neuromuscular clinics after Duchenne muscular dystrophy.

Childhood onset SMA can be classified into three types, distinguished on the basis of clinical severity and age of onset. In type I (Werdnig–Hoffman disease), onset occurs within the first six months of life and children usually die within two years. In type II (intermediate type Dubowitz disease) onset is before 18 months with death occuring after two years. In type III (Kugelberg–Welander disease), the disease has a later onset and milder, chronic cause with affected children achieving ambulation.

At least three genes have been reported to be associated with the SMA type I phenotype on chromosome 5, namely *SMN*, *NAIP* and *p44*. Diagnostic analysis in SMA patients is restricted largely to analysis of the *SMN* gene. The *SMN* gene is present in two copies, one centromeric (*SMNC*) and one telomeric (*SMNT*). The absence of exons 7 and 8 in both copies of the *SMNT* gene is a very reliable diagnostic test for the majority of patients, confirming the clinical diagnosis of SMA. Point mutations have been detected in affected individuals who do not have homozygous deletions. The PCR-based assay used for determining the presence or absence of the SMNT gene is not able to detect individuals who are heterozygous deletion carriers, and a gene dosage method of analysis has been developed to improve carrier detection.

Duchenne and Becker muscular dystrophies

Duchenne muscular dystrophy (DMD) and the milder Becker form (BMD) are X linked recessive disorders causing progressive proximal muscle weakness, associated with elevation of serum creatine kinase levels. Weakness of the diaphragm and intercostal muscles leads to respiratory insufficiency, and involvement of the myocardium causing dilated cardiomyopathy is common.

Both DMD and BMD result from mutations in the gene encoding dystrophin, located at Xp21. The gene is one of the largest identified covering approximately 2.5 megabases of DNA and having 79 exons. Two-thirds of cases are caused by deletion of one or more of the dystrophin exons that cluster in two hot-spots within the gene. Large duplications account for a further 5–10% of cases. The remainder of cases are due to a variety of point mutations.

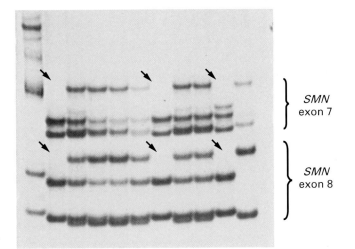

Figure 18.9 Detection of exon 7 and 8 deletions in the *SMN1* gene associated with spinal muscular atrophy can be detected using SSCP analysis. Samples with deletions are indicated by the arrows (courtesy of Dr Andrew Wallace, Regional Genetic Service, St. Mary's Hospital, Manchester)

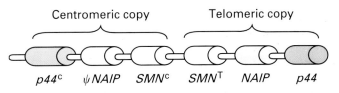

Figure 18.10 The region of chromosome 5 involved in spinal muscular atrophy includes duplications, repetitive regions, truncated genes and pseudogenes making molecular analysis difficult. The suggested genomic organisation of the SMA critical region is shown: *p44* = Subunit of the basal transcription factor TFIIH; *NAIP* = Neuronal apoptosis inhibitory protein gene (Ψ = pseudogene); *SMN* = Survival motor neurone gene. Redrawn from Biros & Forest, *J Med Genet* **36**, 1–8 (1999)

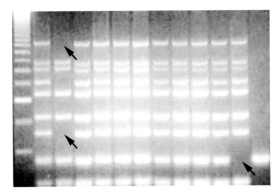

Figure 18.11 Deletion analysis of the dsytrophin gene by multiplex PCR. This analysis simultaneously amplifies exons 43, 45, 47, 48, 50, 51, 52, 53, & 60 with deletions causing loss of bands (arrowed) (courtesy of Dr Simon Ramsden, Regional Genetic Service, St. Mary's Hospital, Manchester)

Since just about all types of mutations can be seen in DMD/BMD cases, a variety of techniques need to be used to carry out a comprehensive molecular analysis. A multiplex PCR approach in which a number of segments of the gene are amplified simultaneously has been developed to rapidly detect deletion of exons in males. Fluorescent dosage analysis can be used to detect deletions and duplications in both affected males and female carriers and chromosomal analysis using fluorescence in situ hybridisation (FISH) techniques will also detect deletions in female carriers. Detecting point mutations is possible with a variety of methods including SSCP analysis, DGGE analysis, and DNA sequencing but is not routine because of the very large number of exons in the gene. In cases where the underlying mutation is unknown, carrier detection and prenatal diagnosis may still be accomplished by linkage analysis with a combination of intragenic DNA markers and markers flanking the gene.

Familial breast cancer

Breast cancer is the commonest cancer seen in young women from developed countries, affecting about 20% of all women who die of cancer. Although the majority of breast cancer cases are sporadic, approximately 5% have an inherited component. The two susceptibility genes identified so far are *BRCA1* and *BRCA2*. The *BRCA1* gene on chromosome 17q21 is involved in 45–50% of inherited breast-only cancer and 75–80% of inherited breast/ovarian cancer. The *BRCA2* gene on chromosome 13q12-13 is involved in approximately 35% of inherited breast-only cancer and 20% of breast/ovarian cancer. In addition, *BRCA2* is involved in a significant proportion of male breast cancer.

Both *BRCA1* and *BRCA2* genes are large, containing 24 and 26 exons respectively. Since being isolated, a considerable number of mutations have been described in both genes – over 250 in *BRCA1* and over 100 in *BRCA2*. Up to 90% of these mutations are predicted to produce a truncated protein. This makes it possible to screen for mutations in the large central exon 11 using the protein truncation test. The remaining exons are generally screened one-by-one using methods such as SSCP/heteroduplex analysis or DNA sequencing. Population-specific founder mutations have been found in eastern European, Ashkenazi Jewish and Icelandic populations. Screening for the common mutation is therefore undertaken as the first step in investigating families from these population groups.

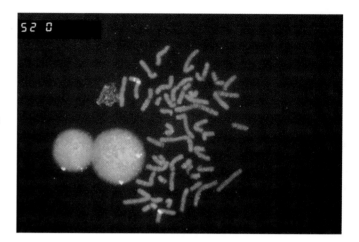

Figure 18.12 Flourescence in situ hybridisation in a female carrier of a Duchenne muscular dystrophy mutation involving deletion of exon 47. Hybridisation with a probe from the centromeric region of the X chromosome identifies both chromosomes. Only one X chromosome shows a flourescent hybridisation signal with a probe corresponding to exon 47, which indicates that the other X chromosome is deleted for this part of the gene (courtesy of Dr Lorraine Gaunt, Regional Genetic Service, St. Mary's Hospital, Manchester)

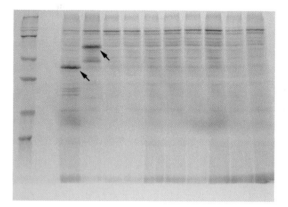

Figure 18.13 Detection of mutations in the *BRCA1* gene that cause premature termination of translation using the protein truncation test. The truncated protein products are shown by the arrows (courtesy of Dr Julie Wu, Regional Genetic Service St Mary's Hospital, Manchester)

19   Treatment of genetic disorders

The prevention of inherited disease by means of genetic and reproductive counselling and prenatal diagnosis is often emphasised. Genetic disorders may, however, be amenable to treatment, either symptomatic or potentially curative. Treatment may range from conventional drug or dietary management and surgery to the future possibility of gene therapy. The level at which therapeutic intervention can be applied is influenced by the state of knowledge about the primary genetic defect, its effect, its interaction with environmental factors, and the way in which these may be modified.

Conventional treatment

Increasing knowledge of the molecular and biochemical basis of genetic disorders will lead to better prospects for therapeutic intervention and even the possibility of prenatal treatment in some disorders. In the future, treatment of common multifactorial disorders may be improved if genotype analysis of affected individuals identifies those who are likely to respond to particular drugs. In most single gene disorders, the primary defect is not yet amenable to specific treatment. Conventional treatment aimed at relieving the symptoms and preventing complications remains important and may require a multidisciplinary approach. Management of Duchenne muscular dystrophy, for example, includes neurological and orthopaedic assessment and treatment, physiotherapy, treatment of chest infections and heart failure, mobility aids, home modifications, appropriate schooling, and support for the family, all of which aim to lessen the burden of the disorder. Lay organisations often provide additional support for the patients and their families. The Muscular Dystrophy Organisation, for example, provides information leaflets, supports research, and employs family care officers who work closely with families and the medical services.

Environmental modification

The effects of some genetic disorders may be minimised by avoiding or reducing exposure to adverse environmental factors. These environmental effects are well recognised in common disorders such as coronary heart disease, and individuals known to be at increased genetic risk should be encouraged to make appropriate lifestyle changes. Single gene disorders may also be influenced by exposure to environmental triggers. Attacks of acute intermittent porphyria can be precipitated by drugs such as anticonvulsants, oestrogens, barbiturates and sulphonamides, and these should be avoided in affected individuals. Attacks of porphyria cutanea tarda are precipitated by oestrogens and alcohol. In individuals with glucose-6-phosphate dehydrogenase deficiency, drugs such as primaquine and dapsone, as well as ingesting fava beans, cause haemolysis.

Exposure to anaesthetic agents may be hazardous in some genetic disorders. Myotonic dystrophy is associated with increased anaesthetic risk and suxamethonium must not be given to people with pseudocholinesterase deficiency. Malignant hyperthermia (MH) is an autosomal dominant condition in which individuals with MH susceptibility, who are otherwise healthy, may develop life-threatening hyperpyrexial

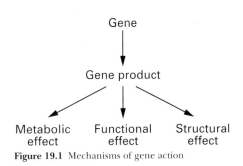

Figure 19.1 Mechanisms of gene action

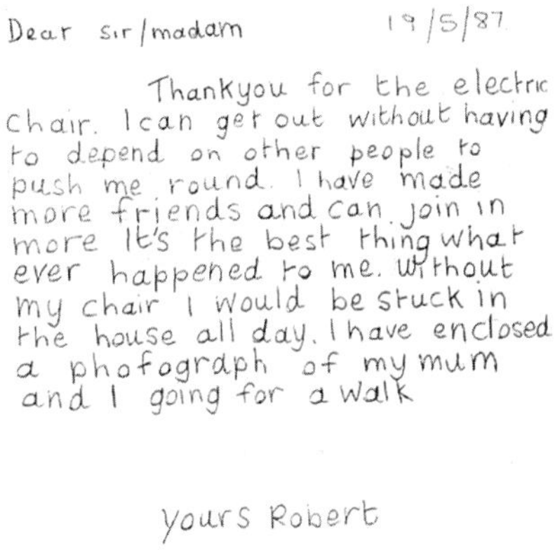

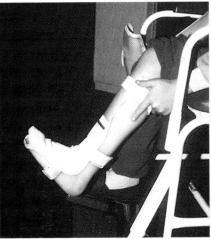

Figure 19.2 Letter written by boy aged 11 with Duchenne muscular dystrophy

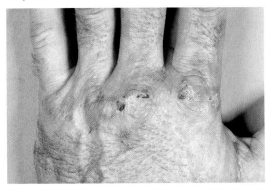

Figure 19.3 Ankle splint used in Duchenne muscular dystrophy (courtesy of Jenny Baker, Clinical Nurse Specialist, Royal Manchester Children's Hospital)

Figure 19.4 Porphyria cutanea tarda (courtesy of Dr Timothy Kingston, Department of Dermatology, Macclesfield General Hospital)

reactions when exposed to a variety of inhalational anaesthetics and muscle relaxants. Relatives with MH susceptibility can be identified by muscle biopsy and in vitro muscle contracture testing. This enables them to ensure that they are not exposed to the triggering agents in any future anaesthetic. It is recommended that susceptible individuals wear a MedicAlert or similar medical talisman containing written information at all times.

Exposure to sunlight precipitates skin fragility and blistering in all the porphyrias except the acute intermittent form. Sunlight should also be avoided in xeroderma pigmentosum (a rare defect of DNA repair) and in oculocutaneous albinism because of the increased risk of skin cancer.

Surgical management

Surgery plays an important role in various genetic disorders. Many primary congenital malformations are amenable to successful surgical correction. The presence of structural abnormalities is often identified by prenatal ultrasound scanning, and this allows arrangements to be made for delivery to take place in a unit with the necessary neonatal surgical facilities when this is likely to be required. In a few instances, birth defects such as posterior urethral valves, may be amenable to prenatal surgical intervention. In some disorders surgery may be required for abnormalities that are secondary to an underlying metabolic disorder. In girls with congenital adrenal hyperplasia, virilisation of the external genitalia is secondary to excess production of androgenic steroids in utero and requires reconstructive surgery. In other disorders, structural complications may occur later, such as the aortic dilatation that may develop in Marfan syndrome. Surgery may also be needed in genetic disorders that predispose to neoplasia, such as the multiple endocrine neoplasia syndromes, where screening family members at risk permits early intervention and improves prognosis. Some women who carry mutations in the *BRCA1* or *BRCA2* breast cancer genes elect to undergo prophylactic mastectomy because of their high risk of developing breast cancer.

Metabolic manipulation

Some inborn errors of metabolism due to enzyme deficiencies can be treated effectively. Although direct replacement of the missing enzyme is not generally possible, enzyme activity can be enhanced in some disorders. For example, phenobarbitone induces hepatic glucuronyl transferase activity and may lower circulating concentrations of unconjugated bilirubin in the Crigler–Najjar syndrome type 2. Vitamins act as cofactors in certain enzymatic reactions and can be effective if given in doses above the usual physiological requirements. For example, homocystinuria may respond to treatment with vitamin B_6, certain types of methylmalonic aciduria to vitamin B_{12}, and multiple carboxylase deficiency to biotin. It may also be possible to stimulate alternate metabolic pathways. For example, thiamine may permit a switch to pyruvate metabolism by means of pyruvate dehydrogenase in pyruvate carboxylase deficiency.

The clinical features of an inborn error of metabolism may be due to accumulation of a substrate that cannot be metabolised. The classical example is phenylketonuria, in which the absence of phenylalanine hydroxylase results in high concentrations of phenylalanine, causing mental retardation, seizures and eczema. The treatment consists of limiting dietary intake of phenylalanine to that essential for normal growth. Galactosaemia is similarly treated by a galactose free diet. In

Figure 19.5 The MedicAlert emblem appearing on bracelets and necklaces. The MedicAlert foundation registered charity website address is http://www.medicalert.co.uk

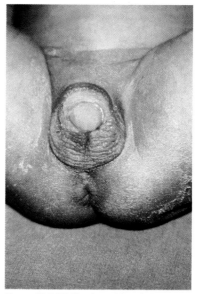

Figure 19.6 Virilisation of female genitalia in congenital adrenal hyperplasia (21 hydroxylase deficiency) (courtesy of Professor Dian Donnai, Regional Genetic Service, St. Mary's Hospital, Manchester)

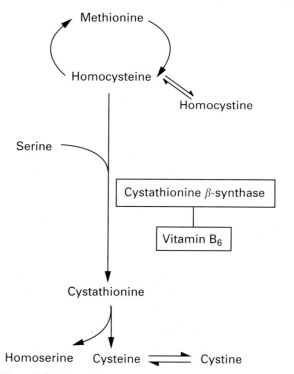

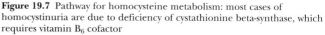

Figure 19.7 Pathway for homocysteine metabolism: most cases of homocystinuria are due to deficiency of cystathionine beta-synthase, which requires vitamin B_6 cofactor

other disorders the harmful substrate may have to be removed by alternative means, such as the chelation of copper with penicillamine in Wilson disease and peritoneal dialysis or haemodialysis in certain disorders of organic acid metabolism. In hyperuricaemia, urate excretion may be enhanced by probenecid or its production inhibited by allopurinol, an inhibitor of xanthine oxidase.

In another group of inborn errors of the metabolism the signs and symptoms are due to deficiency of the end product of a metabolic reaction, and treatment depends on replacing this end product. Defects occurring at different stages in biosynthesis of adrenocortical steroids in the various forms of congenital adrenal hyperplasia are treated by replacing cortisol, alone or together with aldosterone in the salt losing form. Congenital hypothyroidism can similarly be treated with thyroxine replacement. In some disorders, such as oculocutaneous albinism in which a deficiency in melanin production occurs, replacing the end product of the metabolic pathway is, however, not possible.

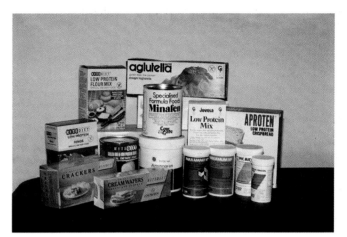

Figure 19.8 Products low in phenylalanine are needed for dietary management of phenylketonuria

Gene product replacement

Gene product replacement therapy is an effective strategy when the deficient gene product is a circulatory peptide or protein. This forms the standard treatment for insulin dependent diabetes mellitus, haemophilia and growth hormone deficiency – conditions that can be treated with systemic injections. This approach is more difficult when the gene product is needed for metabolism within specific tissues such as the central nervous system, where the blood–brain barrier presents an obstacle to systemic replacement.

Genetically engineered gene products are available for clinical use. Recombinant human insulin first became available in 1982. The production of human gene products by recombinant DNA techniques ensures that adequate supplies are available for clinical use and produces products that may be less immunogenic than those extracted from animals. In some cases transgenic animals have been created that produce human gene products as an alternative to cloning in microbial systems.

A potential problem associated with gene product replacement is the initiation of an immunological reaction to the administered protein by the recipient. In haemophilia, the effectiveness of factor VIII injections is greatly reduced in the 10–20% of patients who develop factor VIII antibodies. The efficiency of replacement therapy is, however, demonstrated by the increase in documented life expectancy for haemophiliacs from 11 years in the early 1900s to 60–70 years in 1980. The reduction in life expectancy to 49 years between 1981 and 1990 reflects the transmission of the AIDS virus in blood products during that time period, when 90% of patients requiring repeated treatment became HIV positive. Factor VIII extracts are now highly purified and considered to be free of viral hazard, and recombinant factor VIII has been available since 1994.

An alternative method of replacement is that of organ or cellular transplantation, which aims at providing a permanent functioning source of the missing gene product. This approach has been applied to some inborn errors of metabolism, such as mucopolysaccharidoses, using bone marrow transplantation from matched donors. Again, the blood–brain barrier prevents effective treatment of CNS manifestations of disease. The potential for direct replacement of missing intracellular enzymes in treating inborn errors of metabolism is also being determined experimentally.

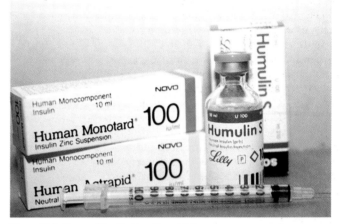

Figure 19.9 Some of the first insulins with human sequence to be prepared biosynthetically or by enzymatic modification of porcine material

Table 19.1 Examples of gene products produced by recombinant techniques for therapeutic use

Product	Disease treated
Alpha interferon	Hairy cell leukaemia
Beta interferon	Multiple sclerosis
Gamma interferon	Chronic granulomatous disease
Factor VIII	Haemophilia A
Factor IX	Haemophilia B
Insulin	Diabetes
Growth hormone	Growth hormone deficiency
Erythropoeitin	Anaemia
DNase	Cystic fibrosis

Box 19.1 Source of cells for replacement therapy

- autotransplantation: use of cells from individual being treated
- allotransplantation: use of cells from another individual
- xenotransplantation: use of animal cells

Gene therapy

The identification of mutations underlying human diseases has led to a better understanding of the pathogenesis of these disorders and an expectation that genetic modification may play a significant role in future treatment strategies. No such treatments are currently available, but many gene therapy trials are underway.

The first clinical trials in humans were initiated in 1990 and since then over 150 have been approved. Most of these have involved genetic manipulation in the therapy of cancer, some have involved infectious diseases or immune system disorders and a few have involved inherited disorders, notably cystic fibrosis. Human trials are all aimed at altering the genetic material and function of somatic cells. Although gene therapy involving germline cells has been successful in animal studies (for example curing thalassaemia in mice) manipulation of human germline cells is not sanctioned because of ethical and safety concerns. So far, results of human gene therapy trials have been disappointing in terms of any long-term therapeutic benefit and many technical obstacles remain to be overcome.

The classical gene therapy approach is to introduce a functioning gene into cells in order to produce a protein product that is missing or defective, or to supply a gene that has a novel function. This type of gene augmentation approach could be appropriate for conditions that are due to deficiency of a particular gene product where the disease process may be reversed without very high levels of gene expression being required. Autosomal recessive and X linked recessive disorders are likely to be the best candidates for this approach since most are due to loss of function mutations leading to deficient or defective gene products. Augmentation gene therapy is not likely to be successful in autosomal dominant disorders, since affected heterozygotes already produce 50% levels of normal gene product from their normal allele. In these cases, gene therapy is not likely to restore gene product production to levels that will have a therapeutic effect. In neoplastic disorders the classical gene therapy approach aims to introduce genes whose products help to kill malignant cells. The genes introduced may produce products that are toxic, act as prodrugs to aid killing of cells by conventionally administered cytotoxic agents, or provoke immune responses against the neoplastic cells.

Genetic manipulation can take place ex vivo or in vivo. In ex vivo experiments and trials, cells are removed and cultured before being manipulated and replaced. This approach is feasible for therapies involving cells such as haemopoetic cells and skin cells that can be easily cultured and transplanted. In in vivo methods, the modifying agents are introduced directly into the individual.

To be effective, augmentation gene therapy requires methods that ensure the safe, efficient and stable introduction of genes into human cells. The production of adequate amounts of gene products in appropriate cells and tissues is needed with appropriate control of gene expression and reliable methods of monitoring therapeutic effects. Before application of gene therapy to humans, in vitro studies are needed together with proof of efficiency and safety in animal models. The possibility of insertional mutagenesis and the dangers of expressing genes in inappropriate tissues need to be considered. There may also be immunological reactions mounted against viral vector material or the gene product itself if this represents a protein that is novel to the individual being treated.

Classical gene augmentation therapy is not suitable for disorders that are due to the production of an abnormal

Table 19.2 Potential application of classical gene therapy approaches

Introduction of normally functioning gene	Correction of phenotype due to absence of gene product
Introduction of toxic gene	Direct cell death in neoplastic or infective diseases
Introduction of prodrug gene	Enhanced cell killing by cytotoxic drugs in neoplastic or infective diseases
Introduction of antigen or cytokine gene	Stimulation of immune response to kill cells in neoplastic or infective diseases

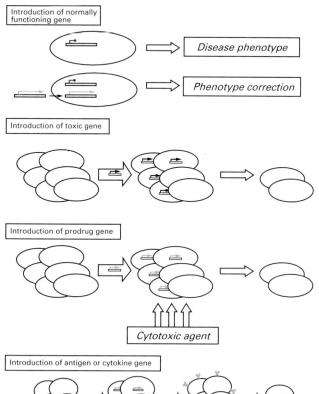

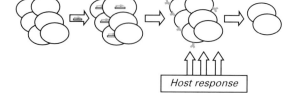

Figure 19.10 Illustrations of gene therapy approaches shown in table 19.2

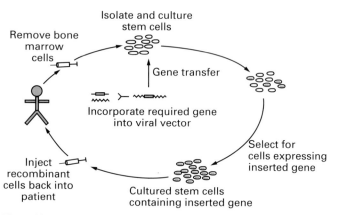

Figure 19.11 Diagram of augmentation gene therapy approach

protein that has a harmful effect because of its altered function. This applies to autosomal dominant disorders where the mutation has a dominant negative effect, producing a protein with a new and detrimental function, as in Huntington disease. Genetic manipulation in this type of disorder requires targeted correction of the gene mutation or the inhibition of production of the abnormal protein product. Several methodologies involving DNA or RNA modification are currently being investigated.

Other approaches to gene therapy include the increased expression of protein isoforms not normally expressed in the affected tissue, or the upregulation of other interacting genes whose products may ameliorate the disease process. In Duchenne muscular dystrophy, for example, it is possible that upregulation of a protein called utrophin, that is related to dystrophin, may have some beneficial effect in slowing the progression of muscle damage.

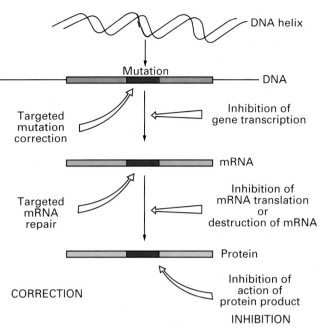

Figure 19.12 New strategies for gene therapy

20 The internet and human genetics

Although many areas of medical science now rely heavily on the internet, human genetics in particular has benefited from its unique ability to provide ready access to information. This is because of the huge quantity of new information that has been generated recently by the Human Genome Project and numerous other research programmes. It is important to remember that not all the information available on the internet is reliable. Anyone with a computer and modem can have their own website and can interpret and disseminate original information in a highly subjective manner. For this reason it is important to use online information that comes from a bonafide source, preferably referenced to original peer-reviewed material. The following section attempts to provide a short guide to websites that may be of relevance to clinical genetics and associated specialties.

Search engines

One of the first problems facing the new internet user is knowing where to start. There are some subject directories providing an overall index rather like a "yellow pages", but most users rely on websites, referred to search engines, that search the internet for them. Not surprisingly, there are a large number of search engines, although each internet service provider will have its preferred website for searching that provides an easy starting point. Website addresses (URLs: uniform resource locators) for a few well-known search engines sites are given below (all are preceded by http://):

- AltaVista™ www.altavista.co.uk
- Lycos™   www.lycos.co.uk
- Google™   www.google.com
- Yahoo™   uk.search.yahoo.com
- GoTo™   www.goto.com
- Cyber411™ www.c4.com (a useful search of search engines)

Human genetics

A useful starting point for general information about human genetics can be found at the British Society for Human Genetics (BSHG) website (www.bshg.org.uk). Along with general information about human genetics and a directory of UK human genetics centres, the BSHG website has links to all major websites involving human genetics. These links include those to the BSHG constituent societies (Clinical Genetics Society, Association of Clinical Cytogeneticists, Clinical Molecular Genetics Society and Association of Genetic Nurses and Counsellors). There are also links to a number of other sites providing useful educational resources, such as online tutorials on genetics.

Finding published literature

The United States National Centre for Biotechnology Information (NCBI) provides a range of invaluable online resources for all types of information on genes and genetics (www.ncbi.nlm.nih.gov/Pubmed). The site provides free access to the PubMed database, which can be rapidly searched for published articles on all aspects of medical research.

Inherited disease databases

The OMIM database (Online Mendelian Inheritance in Man) is a well-established database containing over 11 000 entries on inherited conditions and disease phenotypes. The strength of the site is that entries on each condition rely on peer-reviewed data and are comprehensively referenced, making the information highly reliable. Entries are linked to the PubMed database and to additional resources such as DNA sequence and mapping information. Omim can be accessed through the NCBI website (www.ncbi.nlm.nih.gov/omim) or the UK Human Genome Mapping Project Resource Centre (www.hgmp.mrc.ac.uk/omim).

Information on specific genes

GeneCards (bighost.area.ba.cnr.it/GeneCards or bioinformatics.weizmann. ac.il/cards) is a database of human genes, their products and their involvement in disease. It offers concise information about the functions of all human genes that have an approved symbol, and some others. The human map database can be searched by cytogenetic location, gene or marker name, accession number or the disease name. As with the NCBI databases, the information viewed is now linked to other sites to provide a highly integrated data system known as UDB (Unified Database for human genome mapping).

Mutation databases

The Human Gene Mutation Database (HGMD) is a UK site that provides a rapid method of searching for mutations found in human disease genes and can be accessed using the URL: archive.uwcm.ac.uk/uwcm/mg/hgmd0.html. Entries on each gene are referenced with links provided to the PubMed database. The site also has links to specific gene mutation databases.

Mapping and marker databases

The UK Human Genome Mapping Project (HGMP) funded by the Medical Research Council provides both biological and data resources to the medical research community, with a special emphasis on areas relevant to the Human Genome programme. The Bioinformatics division gives registered users access to a large range of databases and computer programs to aid genomic and proteomic research. The site can be accessed using the URL: www.hgmp.mrc.ac.uk. The number of databases available is huge, and includes analysis services such as BLAST (which searches for sequence similarity between genes), DNA and protein sequences databases, chromosome specific mapping data and databases of genetic markers (e.g. for linkage studies). Other sites providing similar information and links to external sites are the US Human Genome Mapping Project (www.ornl.gov/hgmis) and the European Bioinformatics Institute (www.ebi.ac.uk).

Information on laboratory services and research groups

Locating a UK laboratory that is able to carry out analysis of specific genes can be achieved using the Clinical Molecular Genetics Society website (www.cmgs.org). The site also has links to molecular genetics laboratories throughout the UK. Services offered by molecular genetic laboratories in mainland Europe can be searched using the EDDNAL (European Directory of DNA Laboratories) using the URL: www.eddnal.com

GeneTests™ (www.genetests.org) provides information on genetics clinics, genetic counselling services and genetic testing laboratories in the USA and in other countries. Information is free, although registration is required to use the information. The related site GeneClinics™ (www.geneclinics.org) provides information on specific inherited disorders and the role of genetic testing in the diagnosis, management and genetic counselling of patients with inherited conditions.

Patient organisations

Lay support groups have been established for many genetic conditions. These provide information on specific diseases including research updates and the opportunity for contact between individual families. The larger support groups also organise conferences for families and professionals as well as funding research. In the UK, individual support groups can be contacted through Contact a Family (www.cafamily.org.uk). The Genetic Interest Group (www.gig.org.uk) is an alliance of support groups presenting a unified voice for families in the public arena. Similar groups in the US are the Genetic Alliance (www.geneticalliance.org) and the National Organisation for Rare Disorders (NORD) (www.rarediseases.org).

Websites

General educational resources

MendelWeb (general genetics information)
http://www.netscape.org/Mendel/Web

DNA from the beginning (introductory genetics tutorials)
http://vector.cshl.org/dnaftb

GeneClinics (review articles on several genetic conditions)
http://www.geneclinics.org

Specialised information resources and databases

National Center for Biotechnology Information
(based at NIH, USA)
http://www.ncbi.nlm.nih.gov

On-Line Mendelian Inheritance in Man
http://www.ncbi.nlm.nih.gov/omim or
http://www.hgmp.mrc.ac.uk/omim

UK Human Genome Mapping Project (HGMP)
Resource Centre
http://www.hgmp.mrc.ac.uk

US Human Genome Organisation (HUGO)
Project Information
http://www.ornl.gov/hgmis

HUGO Gene Nomenclature Committee (HGNC)
http://www.gene.ucl.ac.uk/nomenclature

Research Program on Ethical, Legal and Social Implications of
Human Genome Project
http://www.nhgri.nih.gov/ELSI

GeneCards (detailed information about individual
human genes)
http://bighost.area.ba.cnr.it/GeneCards or
http://bioinformatics.weizmann.ac.il/cards

Human Genome Mutation Database (HGMD)
http://archive.uwcm.ac.uk/uwcm/mg/hgmd0.html

European Bioinformatics Institute
http://www.ebi.ac.uk

Familial Cancer Database
http://facd.uicc.org

Genetic societies

British Society for Human Genetics
http://www.bshg.org.uk

Constituent societies
Clinical Genetics Society
http://www.bshg.org.uk/Society/cgs.htm

Association of Genetic Nurses and Counsellors
http://www.agnc.co.uk

Association of Clinical Cytogeneticists
http://www.cytogenetics.org.uk

Clinical Molecular Genetics Society
http://www.cmgs.org.uk

Society for the Study of Inborn Errors of Metabolism
http://www.ssiem.org.uk

Genetical Society
http://www.genetics.org.uk

Irish Society for Human Genetics
http://www.iol.ie/~ishg

European Society of Human Genetics
http://www.eshg.org

American Society of Human Genetics
http://www.faseb.org/genetics/ashg/ashgmenu.htm

Human Genetics Society of Australasia
http://www.hgsa.com.au

American Society of Gene Therapy
http://www.asgt.org

International Federation of Human Genetics Societies
http://www.faseb.org/genetics/ifhgs

UK organisations and committees

Department of Health (Genetics Section)
http://www.doh.gov.uk/genetics.htm

Human Genetics Commission
http://www.hgc.gov.uk

Human Genetics Advisory Commission (now subsumed into
the Human Genetics Commission)
http://www.dti.gov.uk/hgac

Advisory Committee on Genetic Testing (now subsumed into
the Human Genetics Commission)
http://www.doh.gov.uk/genetics/acgt.htm

Gene Therapy Advisory Committee (GTAC)
http://www.doh.gov.uk/genetics/gtac/index.htm

Human Fertilisation and Embryology Authority
http://www.doh.gov.uk/embryo.htm

UK Public Health Genetics Network
http://www.medinfo.cam.ac.uk/phgu

Genetics and Insurance Committee (GAIC)
http://www.doh.gov.uk/genetics/gaic.htm

UK Forum for Genetics & Insurance
http://www.ukfgi.org.uk

Genetics Interest Group
http://www.gig.org.uk

Information about molecular genetic services

Clinical Molecular Genetics Society (lists UK labs offering
molecular tests)
http://www.cmgs.org

GeneTests (US labs offering molecular tests)
http://www.genetests.org

Eddnal (European labs offering molecular tests)
http://www.eddnal.com

Patient information and support networks

Contact-a-Family (UK family support group alliance)
http://www.cafamily.org.uk

Unique. Rare chromosome disorder support group
http://www.rarechromo.org

Antenatal Results and Choices (ARC)
http://www.cafamily.org.uk/Direct/f26.html

European Alliance of Genetic Support Groups (EAGS)
http://www.ghq-ch.com/eags

Genetic Alliance (US patient support group alliance)
http://www.geneticalliance.org

National Organisation for Rare Disorders (NORD) (US)
http://www.rarediseases.org

Glossary

Alleles — Alternative forms of a gene or DNA sequence occurring at the same locus on homologous chromosomes.

Allelic association — The occurrence together of two particular alleles at neighbouring loci on the same chromosome more commonly than would be expected by chance.

Aneuploid — Chromosome constitution with one or more additional or missing chromosomes compared to the full set.

Anticipation — Earlier onset or more severe manifestation of a genetic disorder in successive generations of a family.

Anticodon — Three-base sequence in tRNA that pairs with the three-base codon in mRNA.

Antisense strand (template strand) — DNA strand of a gene used as a template for RNA synthesis during transcription.

Autosome — Any chromosome other than the sex chromosomes.

Autozygosity — Homozygosity for alleles identical by descent in the offspring of consanguineous couples.

Bayesian analysis — Mathematical method for calculating probability of carrier state in mendelian disorders by combining several independent likelihoods.

Candidate gene — A gene identified as being a possible cause of a genetic disease when mutated.

Carrier — A healthy person possessing a mutant gene in heterozygous form: also refers to a person with a balanced chromosomal translocation.

Centromere — The portion of a chromosome joining the two chromatids between the short and long arms.

Chiasma — Visible crossover between homologous chromosomes during prophase stage of meiosis, resulting in exchange of genetic material between the chromosomes.

Chimaera — Presence in a person of two different cell lines derived from fusion of two zygotes.

Chorionic villus sampling (biopsy) — Procedure for obtaining fetally derived chorionic villus material for prenatal diagnosis.

Chromosome — A structure within the nucleus composed of double stranded DNA bearing a linear array of genes that condenses and becomes visible at cell division.

Chromosome painting — Fluorescence labelling of a whole chromosome using multiple probes from a single chromosome.

Clone — An identical copy of the DNA of a cell or whole organism.

Coding DNA — DNA that encodes a mature messenger mRNA.

Codominant — Trait resulting from expression of both alleles at a particular locus in heterozygotes for example, the ABO blood group system.

Codon — Sequence of three adjacent nucleotides in mRNA (and by extension in coding DNA) that specifies an amino acid or translation stop signal.

Complementary DNA (cDNA) — Single stranded DNA synthesized from messenger RNA that contains only coding sequence.

Concordance — Presence of the same trait in both members of a pair of twins.

Congenital — Present from birth.

Consanguinity — Marriage or partnership between two close relatives.

Consultand — The person through whom a family with a genetic disorder is referred to genetic services.

Contiguous gene syndrome — Syndrome caused by deletion of a group of neighbouring genes, some or all of which contribute to the phenotype.

Cytogenetics — The study of normal and abnormal chromosomes.

Deletion — Loss of genetic material (chromosomal or DNA sequence).

Diploid — Normal state of human somatic cells containing two haploid sets of chromosomes (2n).

Discordance — Presence of a trait in only one member of a pair of twins.

Dizygotic twins — Twins produced by the separate fertilization of two different eggs.

DNA — Deoxyribonucleic acid, the molecule constituting genes.

DNA electrophoresis — Separation of DNA restriction fragments by electrophoresis in agarose gel.

DNA fingerprinting — Analysis that detects DNA pattern unique to a given person.

DNA polymerase — Enzyme concerned with synthesis of double stranded DNA from single stranded DNA.

Dominant — Trait expressed in people who are heterozygous for a particular gene.

Duplication — Additional copy of chromosomal material or DNA sequence.

Dysmorphology — Study of malformations arising from abnormal embryogenesis.

Embryo biopsy — Method for preimplantation diagnosis of genetic disorders used in conjunction with in vitro fertilisation.

Empirical risk — Risk of recurrence for multifactorial or polygenic disorders based on family studies.

Epigenetic — Heritable mechanisms not due to changes in DNA sequence, for example methylation patterns.

Eugenics — The use of genetic measures to alter the genetic characteristics of a population.

Euploid — Presence of one or more complete sets of chromosomes with no single chromosomes extra or missing.

Exon — Region of a gene transcribed into messenger RNA.

Fetoscopy — Endoscopic procedure permitting direct visual examination of the fetus.

Fluorescence in situ hybridisation (FISH) — Use of fluorescent nucleic acid probes to detect presence or absence of specific sequences in chromosome preparations or tissue sections.

Frameshift mutation — Mutation that alters the normal reading frame of mRNA by adding or deleting a number of bases that is not a multiple of three.

Gain of function mutation — Mutation that generates novel function of a gene product not just the loss of normal function.

Gamete — Egg or sperm.

Gene — The unit of inheritance, composed of DNA.

Genetic counselling — Process by which information on genetic disorders is given to a family.

Genome — Total DNA carried by a gamete.

Genotype — Genetic constitution of an individual person.

Germline — The cell lineage resulting in formation of eggs or sperm.

Germline (gonadal) mosaicism — Presence of a mutation in some but not all germline somatic cells.

Haploid — Normal state of gametes, containing one set of chromosomes (n).

Haplotype — Particular set of alleles at linked loci on a single chromosome that are inherited together.

Hemizygote — Person having only one copy of a gene in diploid cells (males are hemizygous for most X linked genes).

Heritability — The contribution of genetic as opposed to environmental factors to phenotypic variance.

Heteroplasmy — Presence (usually within single cells) of different mitochrondial DNA variants in an individual.

Heterozygote — Person possessing different alleles at a particular locus on homologous chromosomes.

Holandric — Pattern of inheritance of genes on the Y chromosome.

Homologous chromosomes — Chromosomes that pair at meiosis and contain the same set of gene loci.

Homoplasmy — Presence of identical copies of mitochondrial DNA in the cells of an individual.

Homozygote — Person having two identical alleles at a particular locus on homologous chromosomes.

Hybridisation — Process by which single strands of DNA or RNA with homologous sequences bind together.

Imprinting — Differential expression of a gene dependent on parent of origin.

In-situ hybridization — Hybridisation of a labelled nucleic acid probe directly to DNA or RNA – frequently applied to chromosome preparations or tissue sections.

Interphase — The stage of the nucleus between cell divisions.

Intron — Region of a gene transcribed into messenger RNA but spliced out before translation into protein product.

Isochromosome — Abnormal chromosome composed of two identical arms (p or q).

Karyotype — Description of the chromosomes present in somatic cells.

Kilobase (kb) — 1000 base pairs (bp) of DNA.

Linkage — Term describing genes or DNA sequences situated close together on the same chromosome that tend to be inherited together.

Linkage disequilibrium — See allelic association.

Locus — Site of a specific gene or DNA sequence on a chromosome.

Lyonisation — Process of X chromosome inactivation in cells with more than one X chromosome.

Marker — General term for a biochemical or DNA polymorphism occurring close to a gene, used in gene mapping.

Meiosis — Cell division during gametogenesis resulting in haploid gametes.

Mendelian disorder — Inherited disorder due to a defect in a single gene.

Metaphase — Stage of cell division when chromosomes are contracted and become visible using light microscopy.

Microdeletion — Loss of a very small amount of genetic material from a chromosome, not visible with conventional microscopy.

Microsatellite — Variable run of tandem repeats of a simple DNA sequence widely used for gene mapping in the 1990s.

Mismatch repair — Natural enzymatic process that corrects mis-paired nucleotides in a DNA duplex.

Mis-sense mutation	Nucleotide substitution that results in an amino acid change.
Mitochondria	Cytoplasmic bodies containing mitochondrial DNA and enzymes concerned with energy production.
Mitochondrial inheritance	Exclusively maternal inheritance of mitochondrial DNA.
Mitosis	Cell division occurring in somatic cells resulting in diploid daughter cells.
Modifier gene	Gene whose expression influences the phenotype resulting from mutation at another locus.
Monogenic (unifactorial)	Inheritance controlled by single gene pair
Monosomy	Loss of one of a pair of homologous chromosomes.
Monozygotic twins	Twins derived from a single fertilised egg.
Mosaicism	Presence in a person of two different cell lines derived from a single zygote.
Multifactorial disorder	Disorder caused by interaction of more than one gene plus the effect of environment.
Multiple alleles	Existence of more than two alleles at a particular locus.
Mutation	Alteration to the normal sequence of nucleotides in a gene.
Nondisjunction	Failure of separation of paired chromosomes during cell division.
Obligate carrier	Family member who must be a heterozygous gene carrier, determined from the mode of inheritance and the pattern of affected relatives within the family.
Oncogene	Gene involved in control of cell proliferation that can transform a normal cell into a tumour cell when overactive.
Penetrance	The frequency with which a genotype manifests itself in a given phenotype.
Phenotype	Physical or biochemical characteristics of a person reflecting genetic constitution and environmental influence.
Point mutation	Substitution, insertion or deletion of a single nucleotide in a gene.
Polygenic disorder	Disorder caused by inheritance of several/many susceptibility genes, each with a small effect.
Polymerase chain reaction (PCR)	Method of amplifying specific DNA sequences by repeated cycles of DNA synthesis.
Polymorphism	Genetic characteristic with more than one common form in a population.
Polyploid	Chromosome numbers representing multiples of the haploid set greater than diploid, for example, 3n.
Polysome	Group of ribosomes associated with a particular messenger RNA molecule.
Post-translational modification	Alterations to protein structure after synthesis.
Premutation	A change in DNA that produces no clinical effect, but predisposes to the generation of a pathological mutation.
Proband	Index case through whom a family is identified.
Probe	Labelled DNA or RNA fragment used to detect complementary sequences in DNA or RNA samples.
Promoter	Combination of short DNA sequences that bind RNA polymerase to initiate transcription of a gene.
Pseudogene	Functionless copy of a known gene.
Purine	Nitrogenous base: adenine or guanine.
Pyrimidine	Nitrogenous base: cytosine, thymine or uracil.
Recessive	Trait expressed in people who are homozygous or hemizygous for a particular gene, but not in those who are heterozygous for the gene.
Recombination	Crossing over between homologous chromosomes at meiosis which separates linked loci.
Restriction endonuclease	Enzyme that cleaves double stranded DNA at a specific sequence.
Restriction fragment length polymorphism (RFLP)	Variation in size of DNA fragments produced by restriction endonueclease digestion due to variation in DNA sequence at the enzyme recognition site.
Reverse transcriptase	Enzyme catalysing the synthesis of complementary DNA from messenger RNA.
RNA	Ribonucleic acid, produced by transcription of DNA.
Segregation	Separation of alleles during meiosis so that each gamete contains only one member of each pair of alleles.
Sense strand	DNA strand complementary to the antisense (template) strand, reflecting the transcribed RNA sequence and quoted as the gene sequence.
Sequence tagged sites (STS)	Any unique sequence of DNA for which a specific PCR assay has been designed, enabling rapid detection of the presence or absence of this sequence in any DNA sample.
Sibship	Group of brothers and sisters.
Single nucleotide polymorphism (SNP)	Any polymorphic variation at a single nucleotide position, used for large-scale automated scoring of DNA samples.
Single stranded conformation polymorphism (SSCP)	Commonly used method to screen for point mutations in genes.

Somatic	Involving body cells rather than germline cells.
Southern blotting	Process of transferring DNA fragments from agarose gel onto nitrocellulose filter or nylon membrane.
Splicing	Removal of introns and joining of exons in messenger RNA.
Syndrome	A combination of clinical features forming a recognisable entity.
Telomere	Terminal region of the chromosome arms.
Teratogen	An agent that may damage a developing embryo.
Trait	Recognisable phenotype owing to a genetic character.
Transcription	Production of messenger RNA from DNA sequence in gene.
Translation	Production of protein from messenger RNA sequence.
Translocation	Transfer of chromosomal material between two non-homologous chromosomes.
Trinucleotide repeat	A repeated sequence of three nucleotides that becomes expanded and unstable in a group of genetic disorders.
Triploid	Cells containing three haploid sets of chromosomes (3n).
Trisomy	Cells containing three copies of a particular chromosome (2n + 1).
Tumour supressor gene (TS)	Gene that functions to inhibit or control cell division. Inactivating mutations in TS genes occur in tumours.
Unifactorial (monogenic)	Inheritance controlled by single gene pair.
Uniparental disomy	The inheritance of both copies of a particular chromosome from one parent and none from the other parent.
Uniparental heterodisomy	Inheritance of both chromosomes from a particular homologous pair in the parent.
Uniparental isodisomy	Inheritance of two copies of the same chromosome from a particular homologous pair in the parent.
X inactivation	See lyonisation.
Zygote	The fertilised egg.

Index

Further reading list

Introductory and undergraduate books
Bonthron D, Fitzpatrick D, Porteous M, Trainer A. *Clinical genetics: a case based approach.* London: Saunders, 1998.

Connor JM, Ferguson-Smith MA. *Essential medical genetics.* Oxford: Blackwell, 1997.

Gelehrter TD, Collins FS, Ginsburg D. *Principles of medical genetics.* Baltimore: Williams and Wilkins, 1998.

Mueller RF, Young ID. *Emery's Elements of medical genetics.* Edinburgh: Churchill Livingstone, 1998

Read A. *Medical genetics: an illustrated outline.* London: Gower Medical, 1989.

Thompson M, McInnes J. *Genetics in medicine.* Philadelphia: Saunders, 1998.

Trent RJ. *Molecular medicine: an introductory text.* Edinburgh: Churchill Livingstone, 1997.

Short texts
Harper PS. *Practical genetic counselling.* Oxford: Butterworth Heinemann, 1998.

Weatherall, DJ. *The new genetics and clinical practice.* Oxford: Oxford University Press, 1991.

Young ID. *Introduction to risk calculation in genetic counselling.* Oxford: Oxford University Press, 1991.

King R, Stansfield WD. *A dictionary of genetics.* Oxford: Oxford University Press, 1996.

Snustaad DP, Simmonds MJ. *Principles of genetics.* New York: Wiley, 1997.

Day INM, Humphries SE (eds). *Genetics of common diseases.* Oxford: Bios, 1997.

Ostrer H. *Non-mendelian inheritance in humans.* Oxford: Oxford University Press, 1998.

Reference texts
Rimoin DL, Connor JM, Pyeritz RE, Emery AEH (eds). *Emery and Rimoin's Principles and practice of medical genetics.* Edinburgh: Churchill Livingstone, 2001.

McKusick VA. *Mendelian inheritance in man.* Catalogs of Human Genes and Genetic Disorders 12th edn. Baltimore: Johns Hopkins Press, 1998. (Also available on line).

Vogel F, Motulsky AG. *Human genetics, problems and approaches.* Berlin: Springer, 1996.

Gorlin RJ, Cohen MM, Hennekham RCM. *Syndromes of the head and neck.* Oxford: Oxford University Press, 2001.

Scriver CR, Beaudet AL, Sly WS, Walle D (eds). *Metabolic basis of inherited disease.* New York: McGraw-Hill, 1996.

King RA, Rotter J, Motulsky AG (eds). *The genetics of common disorders.* Oxford: Oxford University Press, 1992.

Khoury MJ, Burke W, Thomson E (eds). *Genetics and public health in the 21st Century.* Oxford: Oxford University Press, 2000.

Specific organ systems
Baraitser M. *The genetics of neurological disorders.* Oxford: Oxford University Press, 1997.

Pulst S-M (ed). *Neurogenetics.* Oxford: Oxford University Press, 2000.

Emery AEH. (ed). *Diagnostic criteria for neuromuscular disorders.* Oxford: Oxford University Press, 1997.

Emery AEH. (ed). *Neuromuscular disorders: clinical and molecular genetics.* New York: Wiley, 1998.

Hagerman RJ, Cronister A (eds). *Fragile X syndrome: diagnosis, treatment and research.* Baltimore: Johns Hopkins, 1996.

Plomin R, Defries JC, McClearn GE, Rutter M. *Behavioural genetics.* New York: Freeman, 1997.

Wynne-Davies K, Hall CM, Apley AG. *Atlas of skeletal dysplasias.* Edinburgh: Churchill Livingstone, 1985.

Sybert VP. *Genetic skin disorders.* New York: Oxford University Press, 1997.

Moss C, Savin J. *Dermatology and the new genetics.* Oxford: Blackwell, 1995.

Traboulski EI. *Genetic diseases and the eye.* Oxford: Oxford University Press, 1998.

Taylor D (ed). *Pediatric ophthalmology.* Oxford: Blackwell Scientific Publications, 1997.

Black GCM. *Genetics for ophthalmologists.* London: ReMEDICA, 2001.

Gorlin RJ, Toriello HV, Cohen MM. *Hereditary hearing loss and its syndromes.* New York: Oxford University Press, 1995.

Cooper DN, Krawczak M. *Venous thrombosis: from genes to clinical medicine.* Oxford: Bios Scientific Publishers, 1997.

Tuddenham EGD, Cooper DN. *The molecular genetics of haemostasis and its inherited disorders.* Oxford: Oxford University Press, 1994.

Cancer genetics
Mitelman P. *Catalog of chromosome aberrations in cancer.* New York: Wiley, 1998. (also available on CD)

Vogelstein B, Kinzler KW. *The genetic basis of human cancer.* New York: McGraw-Hill, 1998.

Hodgson SV, Maher ER. *A practical guide to human cancer genetics.* Cambridge: Cambridge University Press, 1999.

Lalloo FI. *Genetics for oncologists.* London: ReMEDICA, 2001.

Birth defects and dysmorphology
Aase JM. *Diagnositic dysmorphology.* New York: Plenum Medical, 1990.

Cohen MM. *The child with multiple birth defects.* New York: Oxford University Press, 1997.

Jones KL. *Smith's recognisable patterns of human malformation.* Philadelphia: Saunders, 1988.

Baraitser M, Winter R. *A colour atlas of clinical genetics.* London: Wolfe, 1988.

Stevenson RE, Hall JG, Goodman RM, (eds). *Human malformations and related anomalies.* New York, Oxford: Oxford University Press, 1993.

Donnai D, Winter RM, (eds). *Congenital malformation syndromes.* London: Chapman & Hall, 1995.

Winter RM, Knowles SAS, Bieber FR, Baraitser M. *The malformed fetus and stillbirth. A diagnostic approach.* Chichester: Wiley, 1988.

Graham JM. *Smith's recognisable patterns of deformation.* Philadelphia: Saunders, 1998.

Wiedemann H-R, Kunze J. *Clinical Syndromes.* St. Louis: Mosby, 1997.

Hall JG, Froster-Iskenius VG, Allanson JE. *Handbook of normal physical measurements.* Oxford: Oxford University Press, 1989.

Prenatal diagnosis and screening
Abramsky L, Chapple J (eds). *Prenatal diagnosis. The human side.* London: Chapman & Hall, 1994.

Milunsky A. *Genetic disorders and the fetus.* Baltimore: Johns Hopkins, 1998.

Simpson JL, Golbus MS. *Genetics in obstetrics and gynaecology.* Philadelphia: Saunders, 1998.

Wald N, Leck I (eds). *Antenatal and neonatal screening.* Oxford: Oxford University Press, 2000.

Embryology and teratogenesis
Wolpert L, Beddington R, Brockes J, Jessel T, Lawrence P, Meyerowitz E. *Principles of development.* Oxford: Current Biology Ltd & Oxford University Press, 1998.

Moore KL, Persaud TVN. *The developing human. Clinically orientated embryology.* Philadelphia: Saunders, 1998.

Shepard TH. *Catalog of teratogenic agents.* Baltimore: Johns Hopkins, 1998. (also available on CD).

Cytogenetics and chromosomal disorders
Rooney D, Czepulkowski B (eds). *Human cytogenetics: constitutional analysis.* Oxford: Oxford University Press, 2001.

Schinzel A. *Catalogue of unbalanced chromosome aberrations in man.* Berlin: De Gruyter, 1984. (also available on CD).

De Grouchy J, Turleau C. *Clinical atlas of human chromosomes.* New York: Wiley, 1982. (also available on CD).

Gardner RJM, Sutherland GR. *Chromosome abnormalities and genetic counselling.* New York, Oxford: Oxford University Press, 1996.

Molecular genetics
Bridge PJ. *The calculation of genetic risks: worked examples in DNA diagnostics.* Baltimore: Johns Hopkins, 1997.
Strachan T, Read AP. *Human molecular genetics 2.* Oxford: Bios, 1999.

Counselling
Clarke A. (ed). *Genetic counselling. Practice and principles.* London: Routledge, 1994.
Evers-Kiebooms G, Fryns J-P, Cassiman J-J, Van den Berghe H. *Psychosocial aspects of genetic counselling.* New York: Wiley-Liss, 1992.
Baker DL, Schuette JL, Uhlmann WR (eds). *A guide to genetic counselling.* New York: Wiley-Liss, 1998.
Weil J. *Psychosocial genetic counselling.* Oxford: Oxford University Press, 2000.

Social and ethical issues
Harper PS, Clarke A. *Genetics, society and clinical practice.* Oxford: Bios, 1997.
Marteau T, Richards M (eds). *The troubled helix: social and psychological implications of the new genetics.* Cambridge: Cambridge University Press, 1996.
British Medical Association (eds). *Human genetics: choice and responsibility.* Oxford: Oxford University Press, 1998.

Advisory committee reports and consultation documents
Nuffield Council on Bioethics. *Genetic screening – ethical issues.* London: Nuffield Council on Bioethics, 1993.
Working Party of the Clinical Genetics Society. Report on the genetic testing of children. *J Med Genet.* 1994; **31**: 785–97.
Advisory Committee on Genetic Testing. *Code of practice and guidelines on human genetic testing services supplied direct to the public.* London: UK Department of Health, 1997.
Human Genetics Advisory Committee. *The implications of genetic testing for life insurance.* Department of Health, 1997.
Holktzman NA, Watson MS. *Promoting safe and effective genetic testing in the United States.* Washington: NIH, 1997.
Advisory Committee on Genetic Testing. *Genetic testing for late onset disorders.* London: UK Department of Health, 1998.
Human Genetics Advisory Committee. *Cloning issues in reproduction, science and medicine.* Department of Health, 1998.
Gene Therapy Advisory Committee. *Potential use of gene therapy in utero.* Department of Health, 1998.
Human Genetics Advisory Committee. *The implications for genetic testing for employment.* Department of Health, 1999.
Human Fertilisation and Embryology Authority and Advisory Committee on Genetic Testing. *Consultation document on preimplantation genetic diagnosis.* Department of Health, 1999.
Human Genetics Commission. *Whose hands on your genes?* Department of Health, 2000.
Human Genetics Commission. *The use of genetic information in insurance. Interim recommendations.* Department of Health, 2001.

Databases available on CD
Winter RM, Baraitser M. *London dysmorphology database and dysmorphology photo library.* Oxford: Oxford University Press.
Baraitser M, Winter RM. *London neurogenetics database.* Oxford: Oxford University Press.
Bankier A. *POSSUM (dysmorphology database and photo library).* Melbourne, Australia: Murdoch Institute.
Bankier A. *OSSUM (skeletal dysplasia database and photo library).* Melbourne, Australia: Murdoch Institute.
Hall CM, Washbrook J. *Radiological electronic atlas of malformation syndromes and skeletal dysplasias (REAMS).* Oxford: Oxford University Press.
Mitelman P. *Catalog of chromosome aberrations in cancer.* New York: Wiley.
Schinzel A. *Catalogue of unbalanced chromosome aberrations in man.* Berlin: De Gruyte.
De Grouchy J, Turleau C. *Clinical atlas of human chromosomes.* New York: Wiley.
Shepard TH. *Teratogenic agents and risks (TERIS).* Baltimore: Johns Hopkins.

Index

Where not already indexed with a text reference, page numbers in **bold** refer to illustrations; those in *italic* to tabulated or boxed material.

Index

Index

Index